PROJECT MANAGEMENT FUNDAMENTALS

....

A Practical Overview of the PMBOK
Second Edition

George T. Edwards

Project Management Fundamentals – A Practical Overview of the PMBOK - Second Edition

ISBN: 9780979762307

By George T. Edwards

Blue Crystal Press

Atlanta, GA

Editorial Design by: www.graficate.org

Notice of Rights

Notice of Liability

The information given in this book is written in good faith. Even though the content of this book has been researched, some portions could be considered speculative in its own nature. Therefore, the information should be used responsibly and at your own discretion. The information in this book is distributed on an "as is" basis, without warranty. Neither the author, the publisher nor the companies owned by the author shall have any liability to any person or entity with respect to any loss or damage caused by or alleged to be caused directly or indirectly by the information contained in this book.

This publication is designed to provide accurate and authoritative information in regard to the subject matter covered. It is sold with the understanding that the publisher is not engaged in providing professional service.

Trademarks

Blue Crystal Press, Project Management Essentials - A Practical Overview of the PMBOK
Blue Crystal Press, Atlanta, GA
PMP, PMI and PMBOK are registered trademarks of the Project Management Institute

Dedicatory

This book is dedicated to my mother from whom I first learned the concepts of Project Management at the family dinner table and to my father who ran many major projects and taught me valuable lessons in project management.

PREFACE

This books presents the Project Management Body of Knowledge in an easy to read and easy to reference format. It quickly presents what a project manager should do and how it should be done.

By providing a fast and practical overview of the Project Management Body of Knowledge, this book will quickly have the reader applying efficient project management practices and is specifically orientated to people who wish to rapidly grasp what project management is all about.

Every effort has been made to present the concepts in a concise and clear manner, with the result being a book that presents the complex concepts of project management in an easy to use, easy to reference manner.

The book covers the areas of project management knowledge and processes defined by the Project Management Institute. After each process has been described, the book goes on to discuss what all those processes mean in real project environments and which tools to use.

The structure of the book is such that it can be read from beginning to end, but it can also be used as a reference, with readers consulting specific areas of knowledge.

Readers of this book will learn:
- Project management best practices
- How to write and use charters and status reports
- How to plan and schedule a project
- How to manage and control issues and risks
- How to manage and motivate a team
- Project Contracting
- Earned value analysis
- Microsoft Project

There is a section on this book that covers material that is not necessarily used in everyday project management but will be useful when taking the project management certification.

Who Should Read this Book

The first intended audience is people that must learn Project Management techniques "by the book" and in a very short period of time. Perhaps their projects are already starting and they need to know how they are supposed to kick start the project and run it.

The second intended audience is those seeking a PMP certification. This book presents the core material found in the *Guide to Project Management Body of Knowledge* by the Project Management Institute in an easier to read format.

The third audience is those who supervise project managers and need to understand what project management is all about.

Lastly, the fourth audience includes people who participate in projects as contributors, subject matter experts and people with technical expertise who produce the objects to be delivered. While their role by definition is not project management, their careers will be enriched if they learn about project dynamics. Also, they may have to take over some project management responsibilities at some time and, nowadays, employers demand that employees are able to work in a multidisciplinary way. Technical contributors with project management knowledge and experience are sought after by employers. The sections on communications, project estimation and risk management would be very useful for project team members. The summary explanations of the core principles of project management will provide a clear overview of project management.

We believe that there is a great number of people that will benefit from having the project management concepts presented as this book does. As well as being a great reference book in the library of any project team member.

We have also included chapters that contain PMP certification specific materials. The topics covered in that section are areas that a PMP should understand, but the level of detail is more oriented to the PMP certification.

INDEX

Project Management Fundamentals

Figure 1 *-The Hoover Dam was one of the first major projects that used modern project management tools*

1.1   Project Management Overview

Introduction

While Gantt Charts were already being used in the 20's for projects like the Hoover Dam, it was in the late 50's and early 60's that diverse industries started to recognize that the complexity of their projects required specialized management and the formal discipline of project management was born. This happened as projects became more and more complex and the need for a formal project management discipline became more and more clear. The government and private sector saw that projects without formal project management either failed or had cost overruns. Study after study showed that even small projects would have benefited from formal project management leading to the growth of the project management discipline and the number of its practitioners.

Project management recognition has come a long way as more and more enterprises realize the added value that project management brings to a project. It was only a few years ago that project managers had to explain why they were needed in projects, and now it seems that everybody expects that a project cannot be run without a project manager. Project managers have contributed to this trend by professionalizing the occupation and using formal processes to execute and deliver projects. In the past, most project managers were promoted to project management because they were good technically, but now, those positions are going to project managers with formal training. Learning project management techniques has become a must, not only for project managers but also for team members and this book provides the core of what you need to know.

Managing a project can be daunting. Whether planning a bridge, developing a new website or building a business, you need to employ project management techniques to help you succeed. We'll summarize the top seven practices at the heart of good project management in order to help you achieve project success.

Define the Scope and Objectives

As project manager, you must know where you are supposed to go and have a clear understanding of the project objectives. Suppose you need to build a hotel, are you building a motel, a 4 star hotel, or a resort? The earlier you know what kind of hotel you are building the better management you can provide.

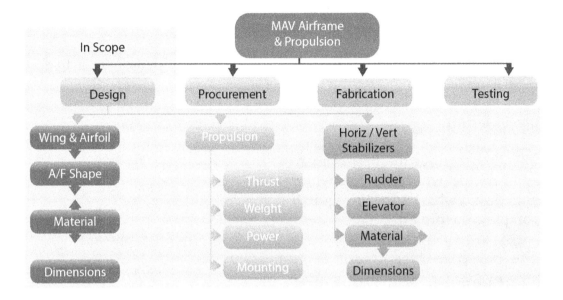

Figure 2 - *The scope for a particular project is many times only one part of bigger project*

Scope defines the boundary of the project. For example, if it is a resort, are the access roads parts of the project? How about beach preparation and cleaning? Deciding what is in or out of the scope will determine the amount of work that needs performing.

Finally, understand who the stakeholders are and what they expect to be delivered. It is important to get their support. Once you have defined the scope and objectives, make sure the stakeholders review and agree with them.

Define the Deliverables

A deliverable is a tangible and verifiable work product.

You must define what will be delivered by the project. If your project is a new purchasing system, then one deliverable might be the training. Therefore, decide what tangible items will be delivered and document them.

You need to make it clear to the sponsors of the project what they will have once the project is completed.

Sponsors and key stakeholders must review the definition of the deliverables and must agree that they accurately reflect what must be delivered.

Project Planning

Planning requires the project manager to decide which people, which resources and what budget is required to complete the project.

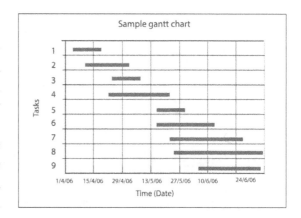

The project manager must define what activities are required to produce the deliverables using techniques such as Work Breakdown Structures. You must estimate the time and effort required for each activity, the dependencies between activities and a realistic schedule to complete them. Involve the project team in estimating how long activities will take. Set the milestones that indicate critical dates during the project. Write this into the project plan. Get the key stakeholders to review and agree to the plan.

Communication

Project plans are useless unless they have been communicated effectively to the project team. Every team member must know their responsibilities. In the same vein, the status of the project must be communicated to all stakeholders and surprises should be avoided. A communication plan specifying what communications will take place, the frequency of those communications and the participants must be created.

Tracking and Reporting Project Progress

Once your project is underway, you must monitor and compare the *actual* progress with the planned progress. You will need progress reports from the project team members. You should record variations between the actual and planned cost, schedule and scope. You should report variations to your manager and key stakeholders and take corrective actions if variations became too large.

You can adjust the plan in many ways to get the project back on track, consequentially you will always end up juggling cost, scope and schedule. If the project manager changes one of these, then one or both of the other elements will inevitably need changing. Juggling these three elements - known as the project triangle - is what typically causes a project manager the most headaches!

Change Management

Stakeholders often change their minds about what must be delivered or the business environment changes after the project starts, so assumptions made at the beginning of the project may no longer be valid. This often means that the scope or the deliverables of the project will have to change. If a project manager accepted all changes into the project, the project would inevitably go over budget, be late or might never be completed.

By managing changes, the project manager can make decisions about whether to incorporate the changes immediately, do it in the future or to reject them all together. This increases the chances of project success because the project manager controls how the changes are incorporated, how resources are allocated, and can plan when and how the changes are made. Not managing changes effectively is often a reason why projects fail.

Risk Management

Risks are events that can adversely affect the successful outcome of the project. I have worked on projects where risks have included the staff lacking the technical skills to perform the work, hardware not being delivered on time, lack of integration and many others. Risks will vary for each project, but the main risks to a project must be identified as soon as possible. Plans must be made to avoid the risk or, if the risk cannot be avoided, to mitigate the risk in order to lessen its impact if it occurs. This is known as risk management.

You cannot manage all risks because there could be too many, and not all risks have the same impact. In this case, it is important to identify all the risks, estimate the likelihood of each risk occurring, estimate its impact on the project and then multiply the two numbers together to obtain the risk factor. High risk factors indicate the severest risks, and you should closely manage the top ten with the highest risk factors. Constantly review risks and look out for new ones as they can occur at any moment.

1.2   Project Management and Its Context

A project is a temporary endeavor undertaken to create a unique product or service. A project is temporary as it has a definite starting and ending point. It is unique as it creates unique deliverables. Finally, a project is progressively elaborated, meaning that it is developed in consecutive steps.

A program is an ongoing concern that groups related projects, which are managed in a coordinated way.

A portfolio is composed of all the programs and projects performed by an organization, and it is created to achieve the organization's objectives.

1.2.1 Project Management

The first challenge of project management is to ensure that a project is delivered within the defined constraints. The second more ambitious challenge is the optimized allocation and integration of the inputs needed to meet those predefined objectives. Therefore, the project is a carefully selected set of activities chosen to use resources (time, money, people, materials, energy, space, provisions, etc.) to meet predefined objectives.

Project management is a combination of disciplines as show in the following graphic that depicts the components of Information Systems Project Management as it can be seen in the following graphic:

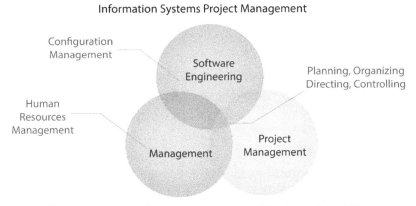

Figure 3 - *Project Management as a combination of disciplines*

1.2.2 Traditional Triple Constraint

Like any human undertaking, projects need to be performed and delivered under certain constraints. Traditionally, these constraints have been listed as scope, time and cost. A further refinement separates product quality or performance from scope and turns quality into a fourth constraint.

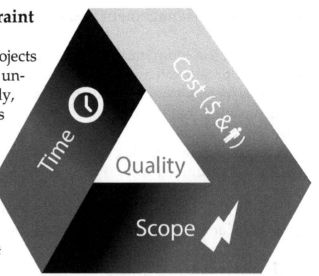

Figure 4- The traditional triple constraint

Table 1 - Project Management Constraints

Time	Cost	Scope
Time required for completing all project tasks versus time available to complete a project.	Cost to develop a project; depends on several variables including (chiefly), labor rates, material rates, equipment, management and profit (when contracting.) This cost is constrained by the budget available for the project.	Requirements specified for the product versus what requirements can actually be accomplished given the time and budget allocated for it.

These three constraints compete with each other. Increased scope typically means increased time and increased cost. A tight time constraint could mean increased costs and reduced scope, and a tight budget could mean increased time and reduced cost.

The discipline of project management provides the tools and techniques which enable the project team (not just the project manager) to organize their work to meet these constraints.

1.2.3 Project Stakeholders

Stakeholders are individuals and organizations impacted by the project and have some level of participation in it. Notice that not all impacted people are stakeholder as competitors can be impacted by a project, but they are not stakeholders.) All stakeholders must be identified and considered for all aspects of the project. Projects can fail if stakeholder interests are not taken into account.

People play different roles in a project and their interactions can be complex, especially if some of the participants have more than one of the essential roles of a project.

These are the basic stakeholder roles:
- Sponsor
- Performing Organization
- Customers / Users
- Functional Managers
- Project Management Office
- Project Contributors
- Project Manager

Sponsor

- Organization (or person) paying for the project
- Defines project scope
- Authorizes project – Issues project charter
- Highest project authority
- Accepts project deliverables

Performing Organization

- Organizational unit charged with executing the project

Customers / Users

People or organizations that will use the deliverables of the project.

Functional Managers

Functional managers are line managers for the project contributors and look after human resource issues for team members. Their influence over the project varies by organizational type (explained in the Project and Organization section). Priorities might conflict with project needs.

Project Management Office

May or not exist in an organization. If it exists, its role can be directive or supportive and its level of authority varies.

The role of the Project Management Office should be to:
- Offer centralized project support
- Define project management processes
- Contribute to project selection
- Be the primary organization for project managers
- Provide project managers to the organization
- Conduct project audits

Project Contributors

Project contributors are project team members that work on the execution of the project, and have diverse functional and technical roles.

Project Manager

With all the emphasis on professionalizing the role of the project manager, the actual role and boundaries still change from organization to organization, but, the essential roles of the project manager are:
- Manage the project following generally recognized good practices of project management
- Determine (in coordination with the project management office and/or

stakeholders) what specific methodologies and processes will be followed for a given project
- Define or influence aspects of project management: charter, schedule, budgets and resources, risk management, etc.
- Report project status to stakeholders
- Take direction from the project sponsor about changes to the project

Given that organizations are in diverse levels of maturity regarding project management, the exact boundaries of the role of a project manager may change not only from organization to organization, but from project to project.

The project manager, to be successful, must have a clear picture of what their role is going to be in a given project. The role must be defined with the sponsor and the executives of the performing organization.

When the project manager has defined limited roles, they could be those of a project coordinator or that of a project expeditor.

A project coordinator does not have resource or financial authority, and their critical role is to identify and report variances from a plan.

1.2.4 Attributes of a Project Manager

Attributes expected from a project manager:
- Leadership
- Communication and coordinating skills
- Attention to detail
- Problem solving skills
- Decision making skills
- Negotiating skills
- Facilitating skills
- Capable of understanding individuals
- Able to cope with ambiguity, setbacks and disappointments
- Ability to view the organization's goals
- Results-oriented
- Can-do attitude

- Politically savvy
- Ability to influence others

While organizations tend to promote technical experts to project leadership positions, being a technical expert does not have to be an attribute of a project manager.

1.2.5 Project Life Cycle

Projects will usually be divided into Project Phases and all the project phases together are also known as the *Project Life Cycle*.

At the conclusion of each project phase, a review of project performance to date is done to determine if the project should continue into its next phase, and to detect and correct errors.

The project life cycle serves to define the beginning and end of a project, and it can be used to link the project to the ongoing operations of the performing organization. It defines:
- What work should be done in each phase
- Who should be involved in each phase

Project life cycle characteristics

- Cost and staffing levels are low at the start, higher toward the end and drop rapidly as the project draws to a conclusion.
- The probability of successfully completing the project is lowest, and hence risk and uncertainty are highest, at the start of the project. The probability of successful completion gets progressively higher as the project continues.
- The ability of a stakeholder to influence the final characteristics of the project's product and final cost of the project is highest at the start and gets progressively lower as the project continues.

The project life cycle often consists of three phases:
- Definition
- Construction
- Closure

Definition Phase

In the definition phase, the scope of the project is established along with the approach to execute it. The project is officially authorized, and a team is formed. The definition stage includes project planning where each task that will need to be executed is identified, assigned and scheduled. It should also include risk analysis and a definition of criteria to be used to accept the project as successful. Planning includes identifying the governance process to be used as well as the chain of command.

Construction Phase

During construction, the tasks are executed. It is the job of the project manager to make sure they are properly executed and that the work products are up to the original standards set during the planning phase. The team must not address the tasks to be executed but resolve issues that arise along the way. Closure Phase

In this last stage, the project manager must ensure that the project is brought to its proper completion. The closure phase is characterized by the writing of a formal project review report containing a formal acceptance of the final product by the client.

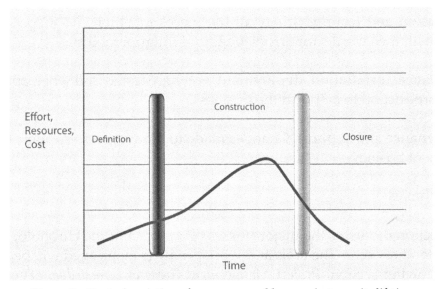

Figure 5 - Typical variation of resources used by a project over its lifetime

1.2.6 Software Development life Cycle

Software development projects have a typical project life cycle which is known as the Software Development Life Cycle (SDLC).

A typical SDLC includes the following steps:

Preliminary Analysis. The objective of this step is to conduct a preliminary analysis, propose alternative solutions, describe costs and benefits and submit a preliminary plan with recommendations.

Systems analysis, requirements definition. Defines project goals into defined functions and operation of the intended application. Analyzes end-user information needs.

Systems design. Describes desired features and operations in detail, including screen layouts, business rules, process diagrams, pseudo code and other documentation.

Development. The computer programs and other objects are built on this phase.

Integration and testing. Brings all the pieces together into a special testing environment, then checks for errors, bugs and interoperability.

Acceptance, installation, deployment. Software tested and when approved, is put into production to help run the business.

Maintenance and support. Changes made to the system to support the evolving nature of business.

Project Phases

This section discusses the major phases of a project in the chronological order that they tend to occur. The tools and work products mentioned will be described in the following section, *Project Management Areas of Knowledge*. These project phases are also the basis for how the PMI organizes process groups.

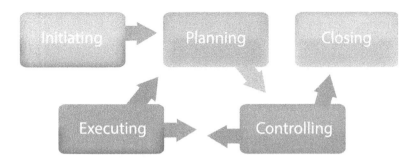

Figure 6 - *Project Phases*

Project Initiation

The projects to be developed are selected from competing projects, based on diverse factors that include cost/benefits analysis and politics.

Project Selection is the process of analyzing the project's potential benefits and its estimated costs, to determine if it should be executed.

Project initiation is the most important step in the formal definition and authorization of a project. In most organizations, it starts with a project charter, but other organizations create the project charter after some planning has been done.

At a minimum, the project initiation should generate a project definition document that identifies the objectives and requirements of the project, the stakeholders, the timeline and resources to be used.

The objectives and requirements of the project as defined in the initiation will be the input for the project planning phase.

Project Planning

Project planning consists of creating a consistent, coherent and comprehensive document that formalizes and details the work to be done.

The project plan is a guide to project execution and project control and includes details not only about the final delivery but how is the project going to be organized and executed.

The most important tool for project planning is the *Work Breakdown Structure* (WBS). It breaks the project into a product- or task-oriented diagram.

Projects are exposed to risks, and those risks should be considered while doing project planning.

Projects can be estimated in a top-down or bottom-up approach. A top-down approach is acceptable in the early stages of planning.

The critical path method is the most important tool used for estimating time-lines.

Human resources actually execute the project, and careful consideration should be given to resource use and availability.

Resource leveling should be used to balance resources and use them more effectively throughout the life of the project. even though that may delay the timeline.

Project Execution

Project execution consists of completing the tasks specified in the project schedule, following the procedures specified in the project plan.

The project manager must assemble the team and manage each individual from the technical and human resources points of view.

During the execution of the project, the project manager's role is to motivate the team members and keep them focused.

Project managers will use situational management to keep the team progressing towards completion.

Project Control

Project control consists of monitoring the project to determine if and how it is following the project plan.

The key tool of project control is change management. Change management is a formal process for identifying change requests, putting them through a decision process and implementing the decision.

Project progress should be reported frequently, and the use of techniques like earned value should be considered to measure project control.

Fast tracking and crashing should be considered to remedy delays.

Project Closing

Project termination is a process that must not be overlooked and consists of bringing a project to a conclusion in an orderly fashion.

At project closeout, the project should be evaluated, and lessons learned should be communicated to the organization.

The team must be dismissed in an orderly fashion.

Even if a project is canceled before completion, the project must be closed properly.

The project closing documentation of one project may be the input for another project.

1.2.7 Projects and Organizations

Enterprise Environmental Factors

Figure 7 - Organization structure and regulations impact projects

Projects are executed within organizations and by organizations. Therefore, these organizations directly influence how the projects are both planned and executed. The influences that the organization exercises over the project are known as Enterprise Environmental Factors (EEF.) EEF are elements that surround and influence project success. The PMI defines EEF as being the inputs related to one third of all project management processes.

Some of the Enterprise Environmental Factors are:
- Company organization
- Company culture
- Industry standards
- Project management maturity
- Laws and regulations
- Stakeholder risk tolerance
- Company infrastructure

Organizational Project Assets

Organizational project assets are project management methods and tools that can be used on project planning and execution. They represent the knowledge acquired from other projects as well as project management tools used by the organization.

As per the PMI, organizational project assets are used as inputs in over half of all project management processes, so it is important to have access to this information.

Examples of organizational project assets are:
- Policies and procedures
- Methodologies
- Forms
- Completed schedules
- Risk data
- Work Breakdown Structures, Tasks Lists
- Project plans and schedules
- Lessons learned

The way a project is organized within an organization influences its operation, and the organization's influence depends on the following:

- Organizational cultures – Which are the values of the organization and how it operates shared values, norms, beliefs, expectations, policies and procedures.
- Authority relationships
- Organizational structure - Organizational structure constraints, the availability of and the terms under which resources become available to the project.

A key organizational component is its project management maturity and by this element alone, organizations can be classified as project–based organizations or non-project based.

In project based organizations, most work is done around projects: teams get formed and dissolved as needed and project contributors flow from one project to the next.

In a non-project-based organization, most work is done by functional areas (departments) and teams are more or less permanent.

While organizations vary in so many different ways, they follow some prototypical patterns as described in the next pages.

1.2.8 Functional Organization

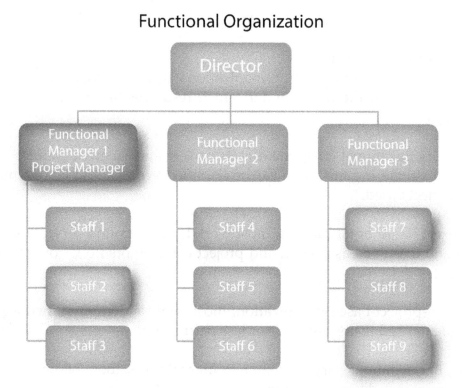

(Shaded boxes represent staff engaged in project activities)

Figure 8 - Functional Organization

The functional organization is the classic hierarchical organization where each employee "belongs" to a department or functional area and reports to one boss from whom they get all direction.

In a functional organization, a functional manager takes project management leadership. Each employee has one clear superior and staff members are grouped by specialty.

The project manager has no formal authority for resources and must rely on informal power structures and their own interpersonal skills to obtain resource commitments from functional managers.

1.2.9 Projectized Organization

(Shaded boxes represent staff engaged in project activities)

Figure 9 - Projectized Organization

In projectized organizations most work is done by teams that are formed to execute the projects and project managers have independence and authority.

The projectized organization is project-oriented and offers better integration among different competencies. People are assigned to work full time for a project and report to the project manager.

A separate, vertical structure is established for each project. All off the project team members report directly and solely to the project manager.

1.2.10 Matrix Organization

A Matrix Organization is a trade-off between project efficiency and business goals. Under this structure, the projects are executed under a functional organization. There are three kinds of matrix organizations: weak, balanced and strong.

Weak Matrix Organization

(Shaded boxes represent staff engaged in project activities)

Figure 10 - Weak Matrix Organization

In a Weak Matrix Organization, the project manager is elected among the staff members that will execute the project.

Vertical functional lines of authority are maintained with a relatively permanent horizontal structure containing managers for various projects.

The balance of power leaning toward the functional manager can cause a project to fall behind, because functional managers are pulling resources away to perform non-project-related tasks.

Balanced Matrix Organization

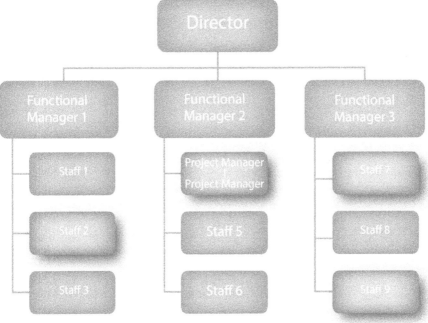

(Shaded boxes represent staff engaged in project activities)

Figure 11 - Balanced Matrix Organization

What makes a balanced matrix organization different from a weak matrix organization is that in the balanced one, the project manager, while being one of the staff members, is also a project manager professional with formal project management training.

Strong Matrix Organization

(Shaded boxes represent staff engaged in project activities)

Figure 12 - Strong Matrix Organization

In the strong matrix organization, there is a project management office that provides project management services to the organization. This is becoming a very common structure.

Figure 13 - The PMO provides project management for the entire organization

1.2.11 Project Management Office

The Project Management Office (PMO) in a business or professional enterprise is the department or group that defines and maintains the standards of processes generally related to project management within the organization. The PMO strives to standardize and introduce economies of repetition in the execution of projects. The PMO is the source of documentation, guidance and metrics in the practice of project management and execution.

1.3   Project Management Areas of Knowledge

1.3.1 Project Management Areas of Knowledge

The project management body of knowledge is actually very vast, and the PMI decided to split the practice of project management into project management areas of knowledge to simplify their study and understanding. The classification made by the PMI has stuck as a good way to study related subject matter with those areas of knowledge being:

- Project Integration Management
- Project Scope Management
- Project Time Management
- Project Cost Management
- Project Quality Management
- Project Human Resource Management
- Project Communications Management
- Project Risk Management
- Procurement Management

Furthermore, PMI divided the project management body of knowledge into management process groups and management processes.

There are five process groups that parallel the project life cycle and they are:

- Initiation
- Planning
- Executing
- Monitoring & Controlling
- Closing

Process Groups Within a Project

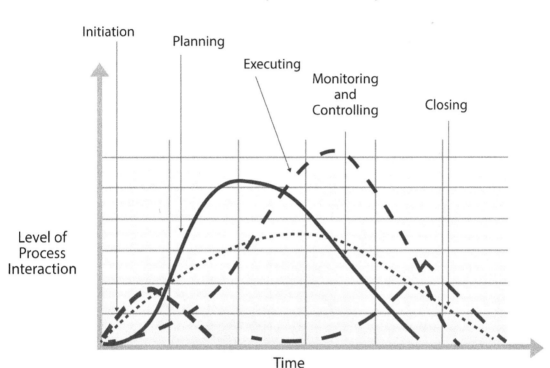

Figure 14 - How the effort for each process group changes over time

The five process groups include forty two project management processes (as per the ANSI/PMI 2008 standard.)

A process is defined as a sequence of steps taken to accomplish a logical unit of work, described independently of any particular implementation that is undertaken by the enterprise to achieve its mission.

Processes are often derived from an analysis of the life cycle of business resources, and may be decomposed into other processes. High-level processes are ongoing in the enterprise. In contrast, lower-level processes have a definable start and end, and are also defined in terms of the input, transformation, output of the process, attributes and relationship types.

1.3.2 Project Management Processes by Process Group

Initiating Process Group	
Knowledge Area	Processes
Integration Management	• Develop Project Charter
Scope Management	
Time Management	
Cost Management	
Quality Management	
HR Management	
Communication Management	
Risk Management	
Procurement Management	

Planning Process Group	
Knowledge Area	Processes
Integration Management	• Develop Project Charter
Scope Management	• Collect Requirements • Define Scope • Create WBS
Time Management	• Define Activities • Sequence Activities • Estimate Activity Resources • Estimate Activity Duration • Develop Schedule
Cost Management	• Estimate Costs • Determine Budget
Quality Management	• Plan Quality
HR Management	• Develop HR plan
Communication Management	• Identify Stakeholders • Plan Communications
Risk Management	• Plan Risk Management • Identify Risks • Perform Qualitative Risk Analysis • Perform Quantitative Risk Analysis • Plan Risk Responses
Procurement Management	• Plan Procurement

Executing Process Group	
Knowledge Area	Processes
Integration Management	• Direct & Manage Execution
Scope Management	
Time Management	
Cost Management	
Quality Management	• Perform Quality Assurance
HR Management	• Acquire Project Team • Develop Project Team
Communication Management	• Distribute Information
Risk Management	
Procurement Management	• Conduct Procurement

Monitoring and Controlling Process Group	
Knowledge Area	Processes
Integration Management	• Monitor & Control Project Work • Perform Integrated Change Control
Scope Management	• Control Scope • Verify Scope
Time Management	• Control Schedule
Cost Management	• Control Costs
Quality Management	• Perform Quality Control
HR Management	• Manage Project Team
Communication Management	• Manage Stakeholders Expectations • Report Performance
Risk Management	• Manage & Control Risks
Procurement Management	• Administer Procurement

Closing Process Group	
Knowledge Area	Processes
Integration Management	• Close Project or Phase
Scope Management	
Time Management	
Cost Management	
Quality Management	
HR Management	
Communication Management	
Risk Management	
Procurement Management	• Close procurement

Table 2 - Project Management Processes

Knowledge Area	Process Groups				
	Initiating	Planning	Executing	Monitoring Controlling	Closing
Integration Management	▪ Develop ▪ Project ▪ Charter	▪ Develop Project Management Plan	▪ Direct & Manage Execution	▪ Monitor & ▪ Control ▪ Project Work ▪ Perform Integrated Change Control	▪ Close Project or Phase
Scope Management		▪ Collect Requirements ▪ Define Scope ▪ Create WBS		▪ Control Scope ▪ Verify Scope	
Time Management		▪ Define Activities ▪ Sequence Activities ▪ Estimate Activity Resources ▪ Estimate Activity Duration ▪ Develop Schedule		▪ Control Schedule	
Cost Management		▪ Estimate Costs ▪ Determine Budget		▪ Control Costs	
Quality Management		▪ Develop HR plan	▪ Perform Quality Assurance	▪ Perform Quality Control	
HR Manage-ment		▪ Identify Stakeholders ▪ Plan Communications	▪ Acquire Project Team ▪ Develop Project Team	▪ Manage Project Team	
Communication Management			▪ Distribute Information	▪ Manage Stakeholders Expectations ▪ Report Performance	
Risk Management		▪ Plan Risk Management ▪ Identify Risks ▪ Perform Qualitative Risk Analysis ▪ Perform Quantitative Risk Analysis ▪ Plan Risk Responses			
Procurement Management		▪ Plan Procurement	▪ Conduct Procurement	▪ Administer Procurement	▪ Close Procurement

The e-book may not display this chart clearly enough to see it. You can download a PDF with the process groups from http://bluecrystalpress.com/process-groups.pdf

1.4   Project Integration Management

1.4.1 Project Integration Management Processes

Project integration management is the area of knowledge about project management that covers the processes required to ensure that the various elements of the project are coordinated.

Table 3 - *Integrative project management processes*

Process Group	Processes	
Initiating	Develop Charter	
Planning	Develop Project Management Plan	
Executing	Direct & Manage Execution	
Monitoring & Controlling	Monitor & Control Project Work	
	Perform Integrated Change Control	
Closing	Close Project or Phase	

Project Integration Management Processes

Develop Project Charter

Consists of developing the formal authorization for a project or phase; including a high-level scope statement for the project.

Develop Project Management Plan

The output of this process is the project management plan that documents all aspects regarding the implementation of the project and includes several subsidiary plans.

Direct and Manage Project Execution

Actions conducive to executing the project management plan to produce the project deliverables.

Monitor and Control Project Work

Monitoring and controlling of all activities of the project to meet the performance objectives.

Perform Integrated Change Control

Process change requests and incorporate them if approved.

Close Project or Phase

Perform closure activities and produce the final project report.

1.4.2 Portfolio Project Management

Before projects can be executed they must be selected from among other projects in the organization.

Project Selection – Two basic approaches:
- Cost/benefit analysis
- Scoring or ranking models

Cost/Benefit Analysis

Traditional approach that attempts to quantify the costs and savings (or revenue) associated with potential projects.

The following techniques can be used for Cost/Benefit Analysis

- Net Present Value (NPV) - A method for evaluating investment proposals in which the value is equal to the present value of future returns, discounted at the cost of capital, minus the present value of the cost of the investment. In general, the higher the NPV, the better the project benefits or returns.

- Life Cycle Costing - The concept of including acquisition, operating and disposal costs when evaluating various alternatives.

- Payback period - This method looks to see how soon the investment will be recovered.

- Return on Investment (ROI) - The rate of return is the financial rate of return considering the expenses on the project minus the pay back amount over a predetermined amount of time. The rate of return is calculated as to how much as a percent the investment is going to generate.

- Internal Rate of Return (IRR) - The internal Rate of return is the minimal return that a company can make by investing in an alternate project or in operative capital. If a project has a ROI greater than IRR, then it is a good investment to make.

NPV is generally the most frequent approach, but life cycle costing is also a very effective way to evaluate projects.

Limitations of Cost/Benefit Analysis

Cost/benefit analysis can be difficult to apply to certain projects due to the project having intangible benefits. (i.e. social and or cultural benefits that are hard to evaluate). The project must be a strategic project that must be done to accomplish certain goals that are considered critical.

Another problem with cost / benefit analysis is that they may fail to account for important qualitative or subjective factors that should be considered in the project selection process.

Scoring and Ranking Models

In order to address the limitations of cost/benefits analysis, enterprises use Scoring and Ranking Models. There is a variety of scoring and ranking models which have been proposed with most models based on a limited number of criteria deemed to be important in selecting projects.

When Ranking, the criteria generally includes quantitative and qualitative factors. These factors are then weighted to reflect their importance in the selection process. Each project is evaluated against all the criteria and a single score is computed per project. The projects with the highest scores are given funding. This way all options are compared against the same factors and compared against each other.

Still, the Scoring and Ranking Models have some limitations:
- Identifying an appropriate set of weights for the selection criteria can be a time-consuming process
- Output of these models is highly dependent on the weights that are assigned to different selection criteria
- The models force us to reduce a multidimensional set of criteria into a single number
- Like traditional cost/benefit analysis, these models focus on one project at a time
- Can use cost minimization or revenue maximization ranking

In some cases it is necessary to verify that a project can be executed and/or that its premises are true. This requires a project in itself. Projects like this are called *feasibility projects*. A feasibility project is a project designed to prove, or disprove, the appropriateness of the technology solution under existing constraints (sometimes called "proof-of-concept" project). A feasibility study is the method and techniques used to estimate technical, cost and resource data to determine potential and practicality of achieving project objectives.

1.4.3 Integrated Change Control

Figure 15 - *Project deliverables as train tracks must be changed in a controlled fashion*

Change is a reality in any project situation. Projects occur over a span of time, and time brings new circumstances and conditions. Projects are also completed by people, and people have new ideas, recognize mistakes and change their minds. Under these circumstances it would be unwise, if not impossible, to proceed with any project without recognizing, accepting and preparing for the possibility of change. Also, a large number of projects fail due to 'scope creep'. Scope creep happens when apparently inconsequential additions to the scope are made. Each of these changes might not bring down a project, but some of them may bring unintended consequences, and their accumulation will drain resources and add time to the project. Lack of change control opens the door to scope creep.

To control the scope of your project, you need to undertake a strict change management process. This process ensures that changes to the project scope, deliverables, timescales or resources are formally defined, evaluated and approved prior to implementation.

The purpose of change control is to prevent downstream work from changing the direction of the project without proper verification and supervision. Change control assures that change is done following a pre-established process, and that change does not derail the project.

Change management starts at the beginning of the project by establishing a change management plan. A change management plan establishes the process to make changes, including who has the authority to make what kind of changes.

Change management should be applied not only to the final product of the project (scope) but also to each work product created during the project. Work product change management starts when a work product is reviewed and approved. At this stage, it is said that the work product is base-lined. If during the course of subsequent work it is determined that the work product needs to be changed, it should be changed through a pre-established process of revision and preapproval.

Change control activities include:
- Revision of requested changes
- Approval process
- Implementation
- Tracking
- Status reporting
- Closure

The project manager does a balancing act regarding change management. In one hand, he or she must deny changes to avoid scope creep but on the other hand must consider incorporating changes whenever possible to maximize customer satisfaction and project performance.

1.4.4 Project Definition Components

Project Charter

The project definition is sometimes called the project charter. A project charter is a document that defines the project and is used by upper management to mark the official creation of a project. The charter is a formal document designed to serve as a contract for the creation of project deliverables. The project charter is a contract between the solution provider and the solution management group. It specifies the overall budget, time constraints, resources and standards. The program or project charter is intended to be a dynamic representation of the project specifications at any point in time. It Describes what needs to be accomplished  in the project at a high level and specifies characteristics of the final deliverable of the project. The charter also establishes the project goal, vision or future state.

The charter provides the manager with the authority to dedicate resources to the project.

The project charter must contain at a minimum:
- The business need that the project is to address
- The product or desired results description
- Deliverables
- Constraints and assumptions
- Timeline
- Budget or resource allocation
- Resources
- Risks
- Key Stakeholders
- Some additional information that can be found in a project charter is:
- Background (programs, environment, history, related projects)
- Strategy - Overall approach to attain the goals of the project
- Assumptions - Factors of the project that are assumed to be true without proof or verification
- Constraints - Any restriction or limitation that could affect the performance of the project
- Acceptance criteria - List of requirements that must be satisfied prior to the customer accepting delivery of the product
- List of stakeholders

Develop Project Management Plan

The project plan development consists of creating a consistent, coherent and comprehensive document that formalizes and details the work to be done.

It is a guide for project execution and project control, and specifies the line of command and communication channels among the different stakeholders.

Project planning includes project scheduling, but they should not be confused with one another.. Project planning defines how the project is going to be executed while project scheduling is about the sequencing of tasks needed to complete the project.

Project Management Plan

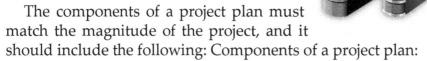

The most important product of the project integration management process is the project management plan.

The components of a project plan must match the magnitude of the project, and it should include the following: Components of a project plan:
- Project charter
- Project management approach or strategy
- Scope statement
- Performance measurement baselines
- Major milestones
- Staff and staffing management plan
- Scope and change management plan
- Risks and risk management plan
- Communications plan
- Quality assurance plan
- Organization structure
- Cost estimates and scheduled dates
- It can also include:
- Work Breakdown Structure (WBS)
- Responsibility chart/assignments
- Network diagram
- Major milestones
- Budget
- Schedule
- Resources

The project management plan is progressively elaborated during the initiating and planning phases of the project. Some of these plans might start as simple paragraphs and then be refined as the projects progresses. For instance, the communication management plan could start as simply as specifying that there will be weekly status reports and there will be weekly status meetings.

Project Plan Creation

Table 4 - Where is the project plan created

Knowledge Area	Processes	Related Management Plan Component Output
Project Integration Management	Develop project management plan	Project Management Plan Schedule Management Plan Cost Management Plan
Project Scope Management	Scope planning	Scope Management Plan
Project Time Management		
Project Cost Management		
Project Quality Management	Quality planning	Quality Management Plan
Project HR Management	HR planning	Staffing Management Plan
Project Communication Management	Communication planning	Communications Management Plan
Project Risk Management	Risk planning	Risk Management Plan
Project Procurement Management	Plan purchases and acquisitions Plan Contracting	Procurement Management Plan Contracting Plan

1.5   Project Scope Management

1.5.1 Project Scope Management Processes

Scope management is the process that ensures that the project includes all the work required to complete the project successfully. It also makes sure that only work required to complete the project is done. In the adjacent figure you can see the representation of many tasks and a few tasks delimited to represent what is in scope and what is not.

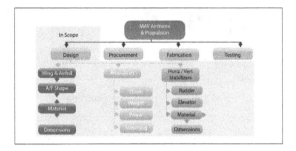

The following processes are part of project scope management:

Process Groups				
Initiating	Planning	Executing	Monitoring Controlling	Closing
	• Collect Requirements • Define Scope • Create WBS		• Control Scope • Verify Scope	

Project Scope Management Processes

Collect Requirements

Consists of gathering and validating the requirements for the project.

Define Scope

Scope definition consists of progressively elaborating on and documenting the project's product or outcome. Scope planning consists of detailing the product and the boundaries of the product that will be created with the project.

Create Work Breakdown Structure

Breaking apart the project deliverables into smaller, more manageable subproducts.

Control Scope
Controlling changes to project scope.

Verify Scope
Verify that the product deliverables meet the project goals.

1.5.2 Project Requirements

Requirements are at the heart of project scope, and you must remember that lack of proper requirements is the major cause of project failure.

Requirements definition guidelines:
* The best approach to define requirements is through feedback from the user and understanding the requirements. This can be done using graphic or physical means, prototyping, simulation, mock screens, etc.
* As important as describing what is included in the project is to mention what is not included in the project
* Stress to the client that what is not in the requirements is not in the project
* State all assumptions and constraints as part of the project definition
* Project requirements must concentrate on the needs, not necessarily on the how
* Establish a procedure to specify changes to the requirements
* *Joint Application* Design and other techniques should be considered

1.5.3 Work Breakdown Structure (WBS)

WBS is a product-oriented division of the components of a project. It is used for defining and organizing the total scope of a project, using a hierarchical diagram.

While the best way to partition a project is to describe the planned outcomes, a WBS is sometimes done to partition the tasks.

Figure 16 - Example of Work Break Down Structure

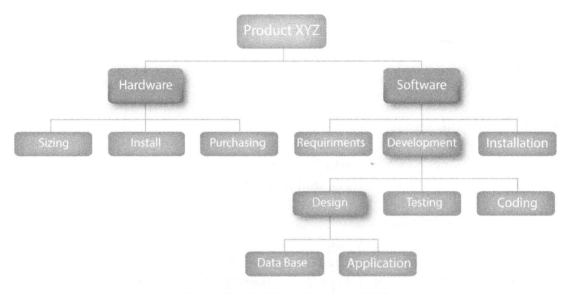

Figure 17 - *Traditional format of WBS*

The WBS is the foundation of project planning and control. It must contain the entire project. If something is not in the WBS, it is not in the project. If the WBS is incomplete, the schedule and budget will be incomplete.

Each level must be defined before a component in a level is broken down further. The breaking down must continue until each component is a work package.

A work package is a piece of work that:
- Can be realistically estimated
- Can be easily assigned
- Can be easily tracked
- Cannot be logically subdivided further
- Can be completed in a manageable amount of time
- Has a clearly definable work product
- Can be completed without interruption

What makes a manageable amount of time depends on the overall project duration, in a yearlong project, work packages could last from one to two weeks. In a month long project, a work package could last from one to two days.

The size of the tasks also depends on the number of total tasks in the project and their interaction.

Work Package - The lowest level of the WBS. This is the level where work is assigned and progress is tracked.

1.5.4 Control Scope

Control Scope consists in all the procedures to be used for all changes that are made to the scope of the project. A key component of Controlling Scope is Scope Change management that is defined as the formal process of recording, analyzing, estimating, tracking and reporting scope changes to the project baseline requirements. Scope Control is tied to Integrated Change Control. Please refer to the section on Integrated Change Control, above, for an additional description of the control scope process. A project manager must be vigilant to prevent scope creep and oftentimes has to "draw a line in the sand" to delimit the scope of the project.

Figure 18 - Project Managers sometimes have to "drawn a line on the sand" to avoid scope creep

Scope creep happens when the scope of a project gets changed a little bit at a time but small changes accumulate and at some point in time, the small changes can sink the project.

1.5.5 Verify Scope

Scope verification occurs at the end of the project, to obtain the user's acceptance of the project. The user will verify that the product matches what was defined in the project plan and what was amended using change control.

1.6   Project Time Management

1.6.1 Project Time Management Processes
Project time management consists of ensuring that the project will be completed in a timely manner.

The following processes are part of the project time management knowledge area:

Knowledge Area	Process Groups				
	Initiating	Planning	Executing	Monitoring Controlling	Closing
Time Management		• Define Activities • Sequence Activities • Estimate Activity Resources • Estimate Activity Duration • Develop Schedule		• Control Schedule	

Project Time Management Processes

Define Activities
The process by which the activities to be performed are defined in enough detail to be assigned. The outputs of this process are activity lists and activity attributes.

Sequence Activities
The discovery or definition of dependencies between tasks. The output is a network diagram and/or an activity predecessor table.

Estimate Activity Resources
This process estimates the types and quantities of resources (including human resources) required to complete the activities as specified.

Estimate Activity Duration
This process determines the time that will be needed to complete each activity with a given set of resources.

Develop Schedule
This process balances all existing inputs to create a schedule that satisfies the project requirements and can be executed with the resources available.

Control Schedule

Activities centered on measuring, reporting and taking corrective action, when needed, about the time used in the execution of activities and the reaching of deadlines.

Time Management Processes Considerations

All these processes interact with each other and they are not necessarily executed in the order listed. There are a number of projects, most notably in the construction industry, where time and cost should be balanced. In those projects where significant expenses are incurred in the overhead, there is an optimum execution calendar time where the overall cost can be minimized.

Defining activities consists of documenting the work to be done and the outputs of each activity. In some cases, the description of the activity is clear if the project follows a known methodology. In other cases, the activity and work products might have to be fully documented.

In some industries, time, resources and cost are more interchangeable than others, so the definition of time and resources ahead of cost is not really always the case. In a number of cases, there are tradeoffs between resources, time and cost.

In this discussion and others, we have used the term activity, the term which has been used for a long time to refer to what PMI calls "Work Package".

1.6.2 Time Planning Tools

Activity

A specific project task, or group of tasks, that requires resources and time to complete. An element of the work performed during the course of a project. Activities are often subdivided into tasks and multiple activities may compose a phase.

Network diagrams

A network diagram is the pictorial representation of the activities and their sequence. The following network diagram shows the tasks and their relationships of precedence.

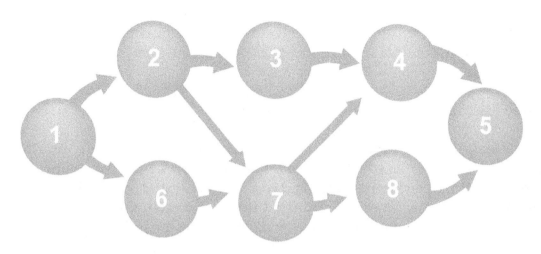

Figure 19 - *Network Diagram*

In this network diagram the tasks are represented by the circles and the sequence or order of execution is represented by the arrows. For example, an arrow
going from task 1 to task 2 means that task 1 must be completed before task 2 can
be completed.

Network diagrams can be done in two ways: activity on arrow diagram and
activity on node diagram. In the activity on arrow, the arrows represent the activities. The idea is that by executing the activity represented by the arrow, the
project moves from one state to another. In the activity on node diagram, the
nodes represent the tasks and the arrows represent the precedence relationship.
While the activity on arrow diagrams appear to be more representative of reality, the activity on node diagrams have proven to be more practical, as a way to
represent and track a project.

Network diagramming is the most important tool used for project planning;
they illustrate the tasks and their relationships in time. The network diagram
shows what precedent activities must be completed before any given activity can
be started, and what following activities can be started when the activity is completed.

Software like MS Project and Primavera use the concept of network diagrams
to define and schedule tasks.

Baseline plan

The initially approved plan to which deviations will be compared as the project proceeds. A work product is something that has been formally approved and that can be changed only through formal change control procedures.

Master schedule

An executive summary which identifies the major components of a project against estimated dates for achievement, particularly those achievements dates which are designated as milestones.

1.6.3 Estimate Activity Resources

Figure 20 - Project Resources can be Human, Machinery or Materials

Resource estimating determines the quality and quantity of resources to be used and the time when they are needed to perform project activities. Resources include people, machinery and materials.

Resource leveling consists of planning the project using the resources most effectively or planning the project using limited resources. Resource leveling could imply longer project duration but less cost and less risk.

Resource planning cannot be done independently of cost estimating, as there will be tradeoffs to consider. Some tradeoffs will even impact the project schedule.

Resource estimating provides an estimation of the needed resource levels to complete project activities.

1.6.4 Gantt Chart

The Gantt chart is a very popular method for displaying the planned execution of activities overtime. It shows overlaps and partially concurrent activities by using horizontal lines to reflect the time required by each activity. The chart, named after Henry Lawrence Gantt, consists of a table of project task information and a bar chart that graphically displays the project schedule to be used in planning and tracking.

	❶	Task Mode ▾	Task Name ▾	Duration ▾	Start ▾	Finish ▾	Predecessors ▾
1		⇥	◢ Build House	45 days	Mon 4/2/12	Mon 6/4/12	
2		⇥	Pour Foundation	1 wk	Mon 4/2/12	Fri 4/6/12	
3		⇥	Build Frame and Walls	2 wks	Mon 4/16/12	Fri 4/27/12	2FS+1 wk
4		⇥	Set up utilities	1 wk	Mon 4/30/12	Fri 5/4/12	3
5		⇥	Put roof	2 wks	Mon 5/7/12	Fri 5/18/12	4
6		⇥	Painting	2 wks	Mon 5/21/12	Fri 6/1/12	5
8		📌	Landscaping	1 wk	Mon 5/21/12	Fri 5/25/12	5
9		📌	House Completed	0 days	Mon 6/4/12	Mon 6/4/12	7,6

Figure 21 - Gantt Charts are critical to project Scheduling

A Gantt chart illustrates a project schedule showing the start and finish dates of the tasks that make up that project.

Milestones are markers used to indicate important points in the timeline; they are usually represented in the Gantt chart as diamonds and represent the culmination or start of a significant task or group of tasks

1.6.5 Calculating the Project Duration

Some key concepts that help to understand schedule planning:
- Working time refers to productive time needed to complete a project. It does not consider breaks, holidays or weekends. For example, it takes 40 hours to write a given reporting program.
- Elapsed time refers to chronological time, includes non-productive time such as lunchtime, and breaks.
- Dependency refers to what tasks need to have been completed to start another task.

Sometimes, dependencies can be complex as when:
- Tasks could need to end together or start together.
- A task would have to start for another to end.

Leads and lags are elapsed time between tasks that share some dependency:
- Lag is the time required between tasks. For example, there is a delay or lag time between ordering equipment and receiving it.
- Lead (or negative lag): When the successor activity must start before the predecessor has been completed.

1.6.6 Estimating Work Time

Estimating work time duration can be a perplexing activity for new project managers and especially for those who have carried out project plans but now have to estimate durations. The truth is that it is difficult to estimate how long a given task is going to take and there is no perfect method to figuring it out. One must rely on judgment and tools that are only going to give an approximation of the real time but it is important to understand that it is better to have an estimate, as imperfect as it can be, than to have no estimate. Another key learning point is that the time estimates can and should be revised once the project is underway. Finally, a good estimate should have a sufficient amount of granularity so it can be explained.

The following work time estimating tools can help estimate times:

Analogous Estimating
Analogous estimating consists in using the actual work time of a previous, similar project as the basis for estimating the work time needed for the current project. This is actually a very good way to do estimating but must be done by an expert who can understand not only the similarities but the differences between the projects.

Parametric Model
The Parametric Model uses project characteristics in a mathematical model to predict project work time. The accuracy of this model is high when the historical information used to implement the model is accurate and parameters used are readily quantifiable

Top-Down Estimating

Based on applying analogous or parametric models to the major components of the project to get a total.

Bottom-Up Estimating

Estimating the time of individual activities, and then summarizing the individual estimates to get a project total. The work time and accuracy is driven by the size and complexity of the individual activity. Requires a WBS at its lowest level.

1.6.7 Critical Path Method

Task Name	Duration	Start	Finish	Predecessors	3/25	4/1	4/8	4/15	4/22	4/29	5/6	5/13	5/20	5/27	6/3
⊿ Build House	45 days	Mon 4/2/12	Fri 6/1/12												
Pour Foundation	1 wk	Mon 4/2/12	Fri 4/6/12												
Build Frame and Walls	2 wks	Mon 4/16/12	Fri 4/27/12	2FS+1 wk											
Set up utilities	1 wk	Mon 4/30/12	Fri 5/4/12	3											
Put roof	2 wks	Mon 5/7/12	Fri 5/18/12	4											
Painting	2 wks	Mon 5/21/12	Fri 6/1/12	5											
House Completed	0 days	Fri 6/1/12	Fri 6/1/12	7,6											6/1

Table 5 - Project Schedule showing the critical path

The critical path method, also known as CPM, is a technique used to find the longest path in a network diagram between the beginning and end of the project, to identify the shortest time in which a project can be completed.

The essential method for using CPM is to construct a network diagram of the project that includes the following:

- All activities required to complete the project
- The time (duration) that each activity will take to complete
- The dependencies between the activities

With these values the CPM methodology calculates the starting and ending times for each activity, determines which activities are critical to the completion of a project (called the critical path) and reveals those activities with "float time" (those that can slide). The critical path is the sequence of project network activities with the longest overall duration, determining the shortest time possible to complete the project. Any delay of an activity on the critical path directly impacts the planned project completion date (in other words, there is no float on the critical path). A project can have several parallel critical paths. An additional parallel path through the network with the total durations shorter than the critical path is called a subcritical or noncritical path.

Float

Float, also called Slack, refers to how much an activity or activities can be delayed and still meet the overall project deadline. There are two types of float that an activity can have:

Free Float or Float

Free Float is the amount of time an activity can be delayed without delaying the early start of an immediate successor.

Total Float

Total Float is the total time that an activity can be delayed without extending the duration of the project.

Project Duration Reduction

Project managers are continually asked to reduce the overall project and there are two recognized ways to reduce the project duration as follows:

Crashing

This technique asks for adding more and more resources to the task to execute it on shorter time but the incremental benefits of adding resources usually keeps diminishing and even can get to the point that more resources actually slow down the task. Cost and schedule tradeoffs are analyzed to obtain the greatest amount of compression for the least incremental cost.

Fast Tracking

With this technique some task that were initially designed to be done one after the other are overlapped and done in parallel. This technique can in fact reduce the overall time but due to tighter coordination, it tends to be more costly and there is a chance that part of the work must be redone.

PERT – Program Evaluation and Review Technique

Another technique for doing schedule planning. It Uses a network diagram. It is used for tasks that have not been done before, for which estimates cannot be done with a reasonable level of exactitude

Uses the following formula to calculate estimated task duration:

PERT Estimated Time =

((Optimistic estimate) + 4 X (Most likely estimate) + (Pessimistic estimate))

$$\overline{}$$

6

Where 'Optimistic estimate' is the time the task would take under the most optimistic circumstances, 'Most Likely estimate' is the best estimate for the task, 'Pessimistic estimate' is the longest time it could take to complete.

Other Scheduling Concepts

Padding - A standard project management tactic used to add extra time or money to estimates to cover for uncertainties and risks of predicting future project activities.

Resource Leveling - The process of shifting resources to even out the workload of team members.

1.6.8 Schedule Control
Schedule control consists of ensuring that the project is being executed as planned and that the tasks are completed according to the timeframes in which they were supposed to be completed.

Schedule control is also concerned with influencing the factors which create schedule changes.

Schedule Control processes ensure that when changes do happen, they are agreed upon and are properly managed and merged into the project.

Tools for schedule control:

- Variance Analysis - Comparing target dates with the actual and forecast start and finish dates
- Earned Value (explained below)

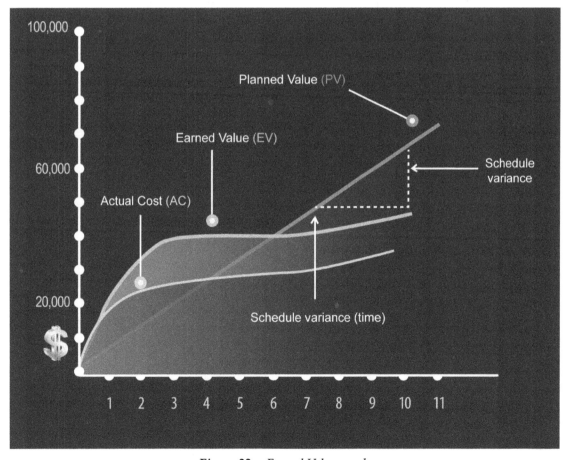

Figure 22 - *Earned Value graph*

Earned Value Analysis (EVA) in its various forms is a widely accepted method of performance measurement. It is used to estimate how a project is doing in terms of schedule and budget.

Earned value is a measure of the value of work performed so far. It uses original estimates and progress-to-date to show whether the actual costs incurred are on budget, and whether the tasks are ahead of or behind the baseline plan.

It can answer the following questions about time and money:
- Where are we now?
- Where are we going?

EVA involves calculating three key values for each activity:
- Planned Value (PV) is the portion of the approved cost estimate planned to be spent on an activity during a given period
- Actual Cost (AC) is the total cost incurred in accomplishing work on the activity during a given period
- Earned Value (EV) is the value of the work actually completed

Example

Planned value, week 1
Task 1: 10 hours at $200 per hour = $2,000
Task 2: 20 hours at $200 per hour = $4,000

Work results at the end of week 1:
Task 1:
Hours worked: 12 => Actual Cost > (12 X $200) = $2,400
Status (percent complete): 100%   (Completed)
Earned Value % complete X   (Total Planned Value)
Earned Value = (100% X $2000)) = $2,000

Task 2:
Hours worked: 5 => Actual Cost > (5 X $200) = $1,000
Status (percent complete): 50%
Earned Value % complete X   (Total Planned Value)
Earned Value = (50% of $4,000) = $2,000

Earned value performance indicators

Cost Variance (CV) = EV – AC
A positive value is good, while a negative value indicates trouble
Cost Performance Index (CPI) = EV / AC
Cost is under control if 0.95 < CPI < 1.10 (Use as a guide, not as a hard rule.)
If the CPI is over 1, the project is progressing at a lower cost than planned.
Schedule Variance (SC) = EV – PV
A positive value is good, while a negative value indicates trouble.
Schedule Performance Index (SPI) = EV / PV
Schedule is under control if 0.95 < SPI < 1.10 (Use as a guide, not as a hard rule.)
If the SPI is over 1 the project is progressing at a faster rate than planned.

Estimate at Completion

Estimate at Completion (EAC) is a forecast of the most likely total project costs based on project performance and risk quantification.

It can be calculated in different ways:
EAC = Current due dates + new estimate for all remaining work
Used when past performance shows that the original estimating assumptions were fundamentally flawed.
EAC = Current due dates + remaining budget
Used when current variances are seen as typical and the project management team's expectations are that similar variances will not occur in the future.
EAC = Actual due dates + remaining budget modified by a performance factor (CPI)
Used when current variances are seen as typical of future variances.

Notes about terminology:
- Planned value is also called Budgeted Cost of Work Scheduled (BCWS).
- Actual cost is also called Actual Cost of Work Performed (ACWP).
- Earned value is also called Budgeted Cost of Work Performed (BCWP).

Earned value must be calculated at the level of cost control and this might be higher than the level of task control. What this means is that tasks are dealt individually when figuring out progress and resources but they are aggregated at a higher level when doing financial analysis.

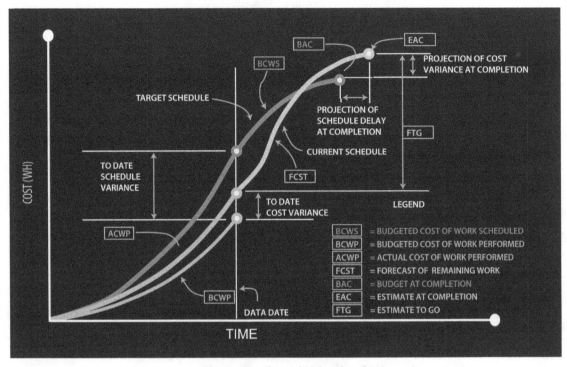

Figure 23 - Earned Value details

1.6.9 Critical Chain Method

The critical chain method is used when resources are limited and the schedule is driven by resources available to the project. It emphasizes resource leveling and uses buffer management instead of earned value management to control the performance of a project. Earned value does not distinguish between progress made in critical path activities versus progress made in other areas.

1.7   Project Cost Management

1.7.1 Project Cost Management Processes

Project cost management consists of assuring that the project is completed within the approved budget and is primarily concerned with the cost of resources used in the project.

The following processes are part of the project cost management knowledge area:

Process Groups				
Initiating	**Planning**	**Executing**	**Monitoring Controlling**	**Closing**
	• Estimate Costs • Determine Budgets		• Control Costs	

Project Cost Management Processes

Estimate Costs
Consists of figuring out an approximation of the costs of all resources to be used in completing a project.

Determine Budget
Budgeting consists of defining cost baselines per the main components of a project. Budgets are used to secure funding for the project and to properly track the cost of the project.

Control Costs
Consists of controlling the expenditure of money and identifying cost variances to keep the project on the planned budget.

1.7.2 Estimate Costs

Key considerations:
• Different costing alternatives must be considered and evaluated

- The estimate costs process must consider whether the cost of additional design work will be offset by expected savings

Estimate cost considers two types of costs:
- Costs associated with carrying out project tasks, such as materials and direct labor costs
- Costs associated with administering the project

Cost estimating tools

Analogous Estimating

- Uses the actual cost of a previous, similar project as the basis for estimating the cost of the current project
- It is used when there is a limited amount of detailed information about the project
- It is the tool used when using expert judgment

Parametric Model

The parametric model consists of using the project characteristics in a mathematical model to predict project costs. The accuracy of this model is high when:
- The historical information used to implement the model is accurate
- Parameters used are readily quantifiable

Scalability

- Top-down Estimating - Based on applying analogous or parametric models to the major components of the project to get a total
- Bottom-up Estimating - Consists in estimating the cost of individual activities, and then summarizing the individual estimates to get a project total. The cost and accuracy is driven from the size and complexity of the individual activity. Requires a WBS at its lowest level

Cost Estimating Risks

- Missing project components (incomplete WBS)

- Incorrect guesses
- Top-down estimates tend to be lower than actual for missing components
- Bottom-up estimates are too costly to make
- Low-balling
- Political pressures

1.7.3 Control Costs

- Monitoring cost performance to detect and understand variances from plan
- Ensuring that all appropriate changes are recorded accurately in the cost baseline
- Prevent incorrect, inappropriate or unauthorized changes from being included in the cost baseline
- Informing appropriate stakeholders of authorized changes
- Acting to bring expected costs within acceptable limits
- Searching out the causes of both positive and negative variances

A key tool for cost control is earned value, which was described in *Project Time Management.*

1.8   Project Quality Management

1.8.1 Quality Management Processes

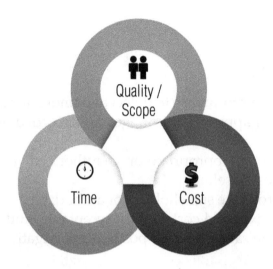

Figure 24 - Quality is one side of the project management triple constraint

The quality knowledge area focuses on the quality of the projects processes and the quality of the deliverables. The quality planning process focuses on identifying standards, while the quality assurance process focuses on project performance. The quality control process focuses on evaluating project results.

Quality is achieved through planning, directing and implementing the actions that are consistent with the concept of "do the right thing right the first time". A dedicated effort of setting standards for the work, understanding the customer's requirements and implementing the requirements in all documentation and actions are needed to infuse quality into projects.

The following processes are part of the project cost management knowledge area:

Process Groups				
Initiating	Planning	Executing	Monitoring Controlling	Closing
	• Estimate Costs • Determine Budget		• Control Costs	

Quality Management Processes

Quality Planning
The process ensures that the project will satisfy the needs for which it was started.

Perform Quality Assurance
Strategic view of quality during the execution of the project undertaken. Takes place over the lifetime of the project.

Perform Quality Control
Operational view of quality during the execution of the project. Takes place at specific times over the lifetime of the project.

Project Quality

Two main types of quality management take place in a project:
- Quality of the project management process
- Quality of the work products

Quality of the Project Management Process

Controlling the quality of the project management process includes the following:
- Verify that the project plan includes the tasks as per project standards
- Verify that the project plan includes the creation of documents as defined by project standards
- Verify that the documents and processes defined in the project plan are actually being executed
- Verify that the planned gates on the project are being performed and that the project actually meets the requirements to pass the gate before continuing

Quality of Work Products

- Quality must be planned and built into the project as opposed to inspecting for quality at the end of the project
- The most important element in keeping the quality of a project high is to follow a consistent methodology for project execution

- The management of quality should start with the project
- Quality not only refers to the quality of the programs created, but also to the technical design, the functional specifications, and even to the project charter
- Use quality to verify conformance to requirements and fitness of use
- Optimal quality is reached at the point where incremental revenue from improvements equals the incremental cost to secure it

Quality Assurance

- Planned and systematic activities implemented within the quality system to provide confidence that the project will satisfy the quality standards and conforms to established requirements
- It consists of applying what has been developed in the quality planning process
- Quality issues must be divided into external and internal quality

Quality Audit

- A structured review of quality management activities
- The objective is to identify lessons learned that can improve organization performance
- Troubles are classified in order to allow future audit
- Audit can often be performed by a third party
- It may be scheduled or random

1.8.2 Quality Concepts

Six Sigma

Six Sigma is a business improvement methodology that systematically improves processes by eliminating errors. Errors are defined as units that are not members of the intended population. The objective of Six Sigma is to deliver high performance, reliability and value to the end customer.

Sigma (the Greek letter σ) is used to represent standard deviation (a measure of variation) of a population, and Six Sigma covers 99.99% of the population. The aim of Six Sigma is to assure that at least 99.99% of the products produced are defect free, specifically only 3.4 defective parts per million.

Deming's 14 Points

W. Edwards Deming redefined quality in the 1950s, and while his studies were orientated toward industrial production, his observations are pertinent to management in general.

He defined his now famous 14 points of quality:

- Create constancy of purpose toward improvement of product and service, with the aim to become competitive and to stay in business, and to provide jobs

- Adopt the new philosophy. We are in a new economic age. Western management must awaken to the challenge, must learn their responsibilities and take on leadership for change

- Cease dependence on inspection to achieve quality. Eliminate the need for inspection on a mass basis by building quality into the product in the first place

- End the practice of awarding business based on price tag. Instead, minimize total cost

- Improve constantly and forever the system of production and service to improve quality and productivity, and thus constantly decrease costs

- Institute training on the job

- Institute leadership. The aim of leadership should be to help people, machines and gadgets to do a better job. Leadership of management is in need of overhaul, as well as leadership of production workers

- Drive out fear so that everyone may work effectively for the company

- Break down barriers between departments. People in research, design, sales and production must work as a team to foresee problems of production and use that may be encountered with the product or service

- Eliminate slogans, exhortations and targets for the work force asking for zero defects and new levels of productivity. These slogans only create adversarial relationships, as the bulk of the causes of low quality and low productivity belong to the system and thus lie beyond the power of the work force

- Eliminate work standards (quotas) on the factory floor. Substitute leadership. Eliminate management by objective. Eliminate management by numbers, numerical goals and substitute leadership

- Remove barriers that rob the hourly worker of his right to pride of workmanship. The responsibility of supervisors must be changed from sheer numbers to quality. Remove barriers that rob people in management and in engineering of their right to pride of workmanship. This means abolishment of the annual merit increase

- Institute a vigorous program of education and self-improvement

- Put everybody in the company to work to accomplish the transformation. The transformation is everyone's job

Pareto Theory

The Pareto Theory establishes that there are only a handful of reasons behind the majority of the problems with quality. The original Pareto theory, which was not related to quality, postulated that 80% of the wealth was owned by 20% of the people. Other people in other areas of expertise found this relationship and in the area of quality it was defined as 80% of problems originate from 20% of the reasons. From here is that we have the 80/20 rule of quality.

80% = 20%
Pareto Principle

Nevertheless, the easy to remember 80/20 rule does not really apply mathematically but it has been found that the general form of the Pareto Theory that can be summed up as "The majority of problems originate from a small number of sources" does hold true for quality control.

Benchmark

A standard figure of merit to which measurements or comparisons may be made.

Cause Effect Graphing

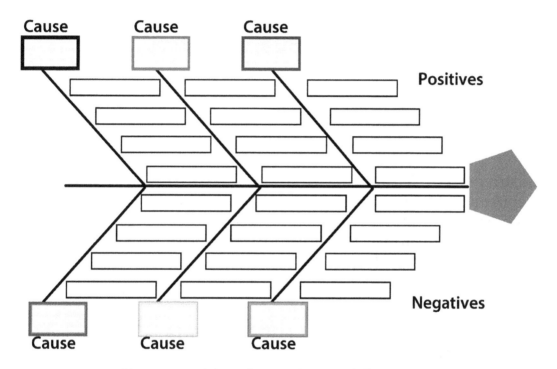

Figure 25 - Fish bone diagram for cause /effect analysis

The cause effect graphic is a test case selection technique in which the input and output domains are partitioned into classes. Analysis is performed to determine which input classes cause which effect. A minimal set of inputs is chosen which will cover the entire effect set.

1.8.3 Software quality concepts

Black Box Testing

Testing based on an analysis of the specification of a piece of software without reference to its internal workings. The goal is to test how well the component conforms to the published requirements for the component.White Box Testing

White Box Testing

Testing based on an analysis of internal workings of a piece of software.

Code Inspection

A formal testing technique whereby the programmer reviews source code with a group of reviewers. The group asks questions analyzing the program logic, analyzing the code with respect to a checklist of historically common programming errors and analyzing its compliance with coding standards.

Code Walkthrough

A formal testing technique where source code is traced by a group with a small set of test cases, while the state of program variables are manually monitored, to analyze the programmer's logic and assumptions.

Concurrency and Performance Testing

Verification that the application can handle the expected concurrent activities and still result in the required response times. Representative data including data values and production data volume must be populated for this testing. This testing can be done in the test environment if the environment is similar to the production environment.

Defect Tracking

During the lifecycle of the project all defects are tracked generally through a tracking tool. The bugs found in testing are reported and the responsible team member is notified to satisfactorily resolve the defect. This method enhances the linkage between tester and developer.

Regression Testing

Rerunning test cases that a program has previously executed correctly in order to detect errors spawned by changes or corrections made during software development and maintenance.

Gate

A gate (or stage-gate) is a step where the merits and progress of the project are evaluated before further progress is allowed. A gate involves a review that often results in a "go/no go" decision for the project.

1.9   Project Human Resource Management

1.9.1 Project Human Resources Processes
The following processes are part of the project cost management knowledge area:

Process Groups				
Initiating	**Planning**	**Executing**	**Monitoring Controlling**	**Closing**
	▪ Develop HR plan	▪Acquire Project Team ▪Develop Project Team	▪ Manage Project Team	

Project Human Resources Processes

Develop Human Resource Plan
Defining project roles and responsibilities and creating the human resources management plan.

Acquire Project Team
Securing the human resources (personnel) needed for a project.

Develop Project Team
Work on improving the individual skills of each team member and improve the collaboration and communication among team members.

Manage Project Team
Supervise team's performance, provide a positive work environment and ensure that team members can perform at optimum levels.

1.9.2 Project Team Management

- Project failures are most often attributed to human failure
- The goal of the project manager is to get the job done
- The job gets done through the people who execute the project
- The project manager must work on the following areas:
- Leadership, inspiration, morale
- Team building
- Team motivation

1.9.3 Motivation Theories

Project managers must motivate the team members and different people have different motivations. What might cheer one person up might not be good for another one. The project manager must identify what motivates each team member and generate the environment needed to have the team motivated.

McGregor's Theory X and Y

Theory X	Theory Y
Workers are inherently self-centered, lazy and lack desire for progress.	Workers are willing and eager to accept responsibility and the work that comes with it.
A worker's most important goal is to obtain money for the work done.	Workers are motivated by things beyond money.
Managers must keep a close eye on employees, as they need to be pushed to work.	Management only needs to create the environment for workers to produce, and they will do their best for the organization.
Top-down authority management works best.	

Maslow's Hierarchy of Needs

Maslow proposed that human beings have a hierarchy of needs. Humans must satisfy certain needs before they can satisfy others, and they can be represented by a pyramid with body (physiological) needs at the bottom

Maslow's Hierarchy:
- **Body (Physiological) Needs** such as air, warmth, food, sleep, stimulation and activity. This need concerns biological balance and stable equilibrium (homeostasis). These needs can be very strong because if deprived over time, the person will die.
- **Security (Safety) Needs** such as living in a safe area away from threats. This level is more likely to be found in children as they have a greater need to feel safe.
- **Social (Love and Belongingness) Needs** such as the love of family and friends.
- **Ego (Self-esteem) Needs** such as healthy pride. Ego needs focus on our need for self-respect and respect from others.
- **Self-Actualization (Fulfillment) Needs** such as purpose, personal growth

and realization of potentials; this is the point where people become fully functional, acting purely on their own volition and having a healthy personality.

1.9.4 Management Styles

Autocratic (directive, dictatorial) management style:
- Directive, top-down approach, does not request nor accept input from subordinates
- May be effective in short term projects where time is absolutely critical
- Could be used to direct non-specialized labor in large projects
- Morale can be affected when this style is used
- Could foster an arbitrary decision-making environment
- Decision-making could lose sight of all factors

Laissez-faire (let do) management style:
- May be effective with highly effective or highly creative individuals that need freedom to do their work
- Morale is high as individuals feel that they are not supervised
- May lead to chaos
- May be inefficient as direction can change multiple times
- Does not work in projects that require quick decision making

Democratic management style:
- Decisions are made with input from team
- Consensus can be reached
- Morale tends to be high as team members are able to influence decisions or at least feel that they are listened to
- Is effective in most project management situations
- Projects may stray from the goals of the organization

Autocratic–Paternalist management style:
- Similar to autocratic, but the well-being of the subordinate is an important input to the decision making

Situational management style:
- Situational management says that a project manager must use any of the management styles as the situation demands

1.10   Project Communications Management

1.10.1 Project Communications Processes

The project communications management area of knowledge focuses on communicating appropriate project information inside the project team and to all stakeholders. It provides a critical link between people, ideas and information at all stages in the project life cycle. Project managers should spend about 60% to 80% of their time communicating. Formal processes aid in decision-making and help to achieve a successful project.

The following processes are part of the project communications management knowledge area:

Process Groups				
Initiating	**Planning**	**Executing**	**Monitoring Controlling**	**Closing**
	• Identify Stakeholders • Plan Communications	•Distribute Information	• Manage Stakeholders Expectations • Report Performance	

Project Communications Processes

Identify Stakeholders

Identify who the formal and informal stakeholders are. Stakeholders are those that impact the project or are going to be impacted by the project.

Plan Communications

Define who will receive information about the project and when.

Distribute Information

Provide information to stakeholders as defined in the communication management plan.

Report Performance

Collect and distribute information about the progress of the project.

Manage Stakeholders' Expectations

Manage information and issue resolution with stakeholders.

1.10.2 Plan Communications

Communications planning is the process of ascertaining the information and communication needs of project stakeholders. It creates the communications management plan.

The communications management plan:
- Is a document that guides project communications
- Is a description of a collection and filing structure for gathering and storing various types of information
- Is a distribution structure describing what information goes to whom, when and how
- Is a format for communicating key project information
- Is a project schedule for producing the information
- Provides access methods for obtaining the information
- Is a method for updating the communications management plans as the project progresses and develops

When planning the communications plan it is important to follow the enterprise standards for knowledge management. Knowledge management is the overall management process to capture, organize, manage and disseminate knowledge in an organization to improve enterprise effectiveness by avoiding mistakes and avoiding the time to relearn needed knowledge. This is specially important for the documentation and dissemination of lessons learned.

1.10.3 Manage Stakeholders Expectations

This process has been set up to emphasize the importance of addressing the stakeholders concerns. Stakeholders must see that their concerns are being considered evenly, specially if the project will have a negative impact in their concerns.

The project manager should actively address stakeholders concerns so those concerns don't become major issues or roadblocks later.

1.10.4 Distribute Information

Information distribution is the process of ensuring the project stakeholders have needed information available in a timely manner. Getting the right information to the right people at the right time and in a useful format is just as important as developing the information in the first place.

1.10.5 Report Performance

Performance reporting is the process of gathering and distributing project performance information including status reporting, progress measurement and forecasting. This process occurs within the monitoring and controlling process group.

- Performance reporting keeps stakeholders informed about how resources are being used to achieve project objectives
- Status reports describe where the project stands at a specific point in time
- Progress reports describe what the project team has accomplished during a certain period of time
- Project forecasting predicts future project status and progress based on past information and trends
- Status review meetings often include performance reporting

1.10.6 Suggestions for Improving Project Communications

- Manage conflicts effectively (See details below)
- Develop better communication skills
- Run effective meetings (See details below)
- Use e-mail effectively
- Use templates for project communications

1.10.7 Conflict Handling Modes

(In preference order)

- Confrontation or problem-solving: directly face a conflict
- Compromise: use a give-and-take approach
- Smoothing: de-emphasize areas of differences and emphasize areas of agreement
- Forcing: the win-lose approach
- Withdrawal: retreat or withdraw from an actual or potential disagreement

1.10.8 Running Effective Meetings

General guidelines for running effective meetings:
- Determine if a meeting can be avoided
- Define the purpose and intended outcome of the meeting
- Determine who should attend the meeting
- Provide an agenda to participants before the meeting
- Prepare handouts, visual aids and make logistical arrangements ahead of time
- Run the meeting professionally
- Build relationships

1.11   Project Risk Management

1.11.1 Risk Management Processes

A risk is an uncertain event that, if it occurs, has negative effect on a project objective. According to Risk Theory, there are also possible risks that if occur, can have a positive effect in a project but this are highly unusual.

Risk management is the systematic process of identifying, analyzing and responding to project risk.

The following processes are part of the project risk management knowledge area:

Process Groups				
Initiating	Planning	Executing	Monitoring Controlling	Closing
	• Plan Risk Management • Identify Risks • Perform Qualitative Risk Analysis • Perform Quantitative Risk Analysis • Plan Risk Responses		• Manage & Control Risks	

Risk Management Processes

Plan Risk Management
Decide how to approach and plan the risk management for the project.

Identify Risks
Determine which risks might affect the project and document their characteristics.

Perform Qualitative Risk Analysis
Subjective analysis for occurrence, impact and criticality.

Perform Quantitative Risk Analysis
Numerical analysis for occurrence, impact and criticality.

Manage and Control Risk

Keep track of risks and plans. Identify new risks and ensure execution of risk management and response.

Important Risk Considerations

Risk management minimizes the probability of adverse events and their consequences for project objectives; conversely, risk management also looks to maximize the probability and consequences of positive events that may happen to the project

The risk management process must ensure that the level, type and visibility of risk management are commensurate with both the risk and importance of the project to the organization. (The same risk will have a different impact on different projects.)

Risks can be initiated by a 'risk trigger' that is an indication that a risk has occurred or is about to occur.

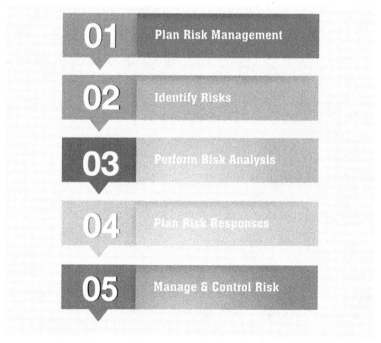

Figure 26 - Risk Management Sequence of processes

1.11.2 Plan Risk Management

The risk management plan is how risk identification, qualitative and quantitative analysis, response planning, monitoring and control will be structured and performed during the project life cycle. It may include:

Roles and Responsibilities
Defines the lead, support, and risk management membership for each type of action in the risk management plan.

Reporting Formats
Describes how the results of the risk management processes will be documented, analyzed and communicated to the project team, stakeholders and sponsors.

Tracking
Documents how all facets of risk activities will be recorded for the benefit of the current project, future needs and lessons learnt.

1.11.3 Risk Identification Tools

There are many risk identification tools. Some of the most critical ones are:
- Brainstorming
- Delphi Technique – The same questions are asked individually to different experts, to get an unbiased opinion
- Interviewing
- Strengths, Weaknesses, Opportunities and Threats (SWOT) Analysis
- External Risks
- Checklist created from similar projects

1.11.4 Perform Qualitative Risk Analysis

Qualitative risk analysis is the process of assessing the impact and likelihood of identified risks. It prioritizes risks according to their potential effect on project objectives. The probability and consequences of the risks are evaluated using established qualitative–analysis methods and tools.

Perform Quantitative Risk Analysis

Quantitative risk analysis looks to quantify the probability of each risk and its consequence on project objectives.

The quantitative risk analysis will provide a chart like the one below where risks are colored white, yellow, orange and red. In the printed version book they appear are shades of gray where the higher the impact and the higher the likelihood of a risk the darker the cell will be.

	Probability of Occurrence			
	Frequent	Occasional	Remote	Unlikely
Very Unlikely (A)	Extreme Risk (9)	Extreme Risk (8)	High Risk (7)	* High Risk (6)
Unlikely (B)	Extreme Risk (8)	High Risk (7)	Medium Risk (5)	* Medium Risk (5)
Possible (C)	High Risk (7)	Medium Risk (5)	Medium Risk (4)	* Low Risk (3)
Likely (D)	Medium Risk (4)	Low Risk (3)	Low Risk (2)	* Low Risk (3)

1.11.5 Plan Risk Responses

Risk response planning is the process of developing options and determining actions to enhance opportunities and reduce threats to the project's objectives. It includes contingency planning that consists of setting up alternative plans of action if the project does not proceed according to plan or if the expected results are not achieved. These alternative plans can be used to ensure the project's success if specified risk events occur.

Mitigation

Take steps to reduce the probability of the occurrence and the impact of the risk if it happens. It may take the form of implementing a new course of action that will reduce the problem (e.g. conducting engineering tests).

Avoidance

Eliminate the risk by changing the project plan.

It consists of changing the project plan to eliminate the risk or to protect the project objective from its impact.
- Reducing scope to avoid high-risk activities
- Adding resources or time

Transference

The risk is transferred to another entity and gives the other entity the ownership of the risk. It involves the payment of a risk premium and does not actually eliminate the risk, just transfers it. A typical transfer situation happens when the project owner buys insurance for an specific risk or hires a third party to execute a portion of the project where the risk, if it were to occur, would have impact.

Acceptance

If the response is accepted, the team may develop a contingency plan, so if the risk actually happens, there is a way to mitigate its impact. The impact could be so low (or so high) that no response is planned ahead of the occurrence.

1.11.6 Other Risk Definitions

- **Residual Risks** - The risk that still exists when the primary risk has been mitigated or transferred.
- **Secondary Risk** - A risk that is generated when an action is taken to address an existing risk.

Risk Response Plan

The risk response plan includes:
- Identifying risks and their possible causes
- Defining risk owners and assigned responsibilities
- Results from qualitative and quantitative risk analysis processes
- Agreed responses including avoidance, transference, mitigation and acceptance

- Residual risk
- Specific actions to implement the chosen response strategy
- Budget and time for response strategy
- Contingency plan and fallback plan

Risk Monitoring

- Keeping track of the identified risks, monitoring residual risks and identifying new risks
- Ensuring the execution of the risk plan
- Evaluating the risk plan's effectiveness in reducing risk
- It is an ongoing process for the life of the project

1.12   Procurement Management

1.12.1 Procurement Management Processes

Procurement management deals with sourcing and buying products and services needed for the project. Purchases can be the largest item in the budget of some projects, especially those involving fabricating large assets.

The following processes are part of the project procurement management knowledge area:

Process Groups				
Initiating	**Planning**	**Executing**	**Monitoring Controlling**	**Closing**
	▪ Plan Procurement	▪ Conduct Procurement	▪ Administer Procurement	▪ Close Procurement

Procurement Management Processes

Plan Procurement

Consists of defining products and services to purchase, identifying sellers and prepare to engage sellers.

Conduct Procurement

Consists of obtaining information, quotations, bids and proposals for products and services. Determining the best offer and negotiate contracts

Administer Procurement

Supervise the execution of the contract, manage vendor relationships, evaluate vendor performance.

Close Procurement

Settling and closing contracts.

1.12.2 The Need for Contracting

Companies have several reasons to contract work to outside firms. In some cases the work to do is just too big for the internal resources, or it could be that the job requires specialized expertise or equipment. Projects are, after all, one-time events that an executing entity may not have the expertise to do and may not need to do again. In those cases, it is better to have the work, or part of it, done by a specialized firm.

Contracts are regulated by laws which are enforced through the legal system. Although the project manager is involved in most aspects of the contract, companies involve lawyers in order to review contracts signed with outside entities. The project manager can participate in all aspects of the contracting, and is especially involved in the contract's administration. The project manager must understand the basic legalities involved in contracting and must know the obligations that each party assumes when a specific contract is signed.

1.12.3 Requirements for a Legal Contract

Contracts must meet several requirements in order to be considered legal and enforceable:

An offer - What one party is offering to do for the other party.

Consideration - The value that is being paid for the work to be done. Most times, this is money, but it could be some valuable good or service.

Acceptance - Both parties must agree to the contract.

Legal Capacity - Both parties must be competent to enter into the contractual agreement. The signing parties must not be minors under the laws of the legal jurisdiction and should be legally permitted to sign a contract. They also must have the legal power to enter into the agreement; this particularly pertains to people representing an outside interest, such as a company or third party.

Legal Purposes - Contracts cannot be written to bind a party to do an illegal act and one cannot have a contract for illegal purposes.

1.12.4 Solicitation Tools

Advertising
Existing lists of potential sellers may often be expanded by placing advertisements in newspapers or in specialty publications.

Bidder Conferences
Meetings with prospective sellers prior to the preparation of proposals. All prospective sellers must have a clear, common understanding of the procurement.

Proposals
Seller-prepared documents that describe the seller's ability to provide the requested product. Proposals may be supplemented with an oral presentation.

Source Selection
- The receipt of bids or proposals and the application of the evaluation criteria to select a provider
- Price may be the primary determinant for an off-the-shelf item
- Proposals are separated into technical and commercial sections
- Multiple sources may be required for critical products

1.12.5 Administer Procurement

The process of ensuring that the seller's performance meets contract requirements. Must make sure that all providers work in harmony and towards a single goal

The project management team must be aware of the legal implications of actions taken during a project

1.12.6 Contract Types

Lump Sum Contract or Fixed Bid

With this kind of contract, the contractor agrees to do a described and specified project for a fixed price. A fixed fee or lump sum contract is suitable if the scope and schedule of the project is sufficiently defined to allow the contractor to estimate project costs.

Time and Materials

Under this contract, the contractor gets paid an hourly rate and is paid for expenses incurred. Used for consulting engagements and labor contracts

Unit Price Contract

This kind of contract is based on estimated quantities of items included in the project and their unit prices. The final price of the project is dependent on the quantities needed to carry out the work.

In general, this contract is only suitable for construction and supplier projects where the different types of items, but not their quantities, can be accurately identified in the contract documents.

Cost-Plus Contract

This is a contract agreement wherein the purchaser agrees to pay the cost of all labor and materials plus an amount for contractor overheads and profit (usually as a percentage of labor and materials cost).

- Cost-plus Contracts can be of at least 3 forms:
- Cost-Plus Fixed Fee - The seller gets its costs paid plus a fixed amount
- Cost-Plus Percentage of Costs - The seller gets its costs paid plus a percentage of the costs
- Cost-Plus Incentive Fee – The seller gets paid its costs plus a fee based on some level of performance that is usually based on time or in not exceeding a budget, or both

Incentive Contracts

Compensation is based on the contracting performance according to an agreed target - budget, schedule and/or quality.

1.12.7 More Contracting Concepts

Acceptance Criteria: The list of requirements that must be satisfied prior to the customer accepting delivery of the product.

Document of Understanding: A formal agreement between two parties that may be a contract on its own, or may be the basis for a more detailed contract.

Service Level Agreement (SLA): A SLA is a contract between a service provider and a customer that specifies, usually in measurable terms, what services the service provider will furnish.

Statement of Work (SOW): An integrated set of task descriptions, goal descriptions, risks, and assumptions that describe a specific task to execute.

Key Project Management Artifacts

2.1 Charter

This is a template of a Project Charter. The instructions are italicized text within gray shaded boxes and should not be included in the final document.

Project Charter

Project Name: _____   Sponsors: _____

Enter the name of the project and the name of the entity that is sponsoring the project or is the main beneficiary.

Revision History

Version	Date	Summary of Changes	Revision Marks

Revision History: Enter the version, date and changes done.

Approvals

Required Approvals: Sponsor, stakeholders and project manager.

The following have approved this document.

Name	Title & Department	Approval Date	Signature

Introduction

Project Description:
Business Problem:
Business Value:

Provide a description of the project in terms of a business need that the project will address. The intended outcome and an explanation of why those outcomes are desirable. Also, describe the business value this project will deliver.

Stakeholders
Ownership & Accountability

Identify all of the stakeholders and users of the system and ensure that the stakeholder community adequately represents them.

Owner	Accountability

Project Scope
Objectives

Provide an overall statement summarizing, at the highest level, the objectives that the project will achieve. Answer the question, "What is the problem we are trying to solve?" The statement should communicate the intent and the importance of the project to all concerned.

Project Dependencies

Identify the projects and/or activities that may affect the successful delivery of this project.

System or Product Features

List and briefly describe the key product features. These features are the critical capabilities of the system that are required to deliver benefits to the users. A Business Requirements Document will contain the full set of requirements for this project.

ID	Feature Description

Major Milestones, Deliverables & Timeline

Identify major milestones and deliverables (i.e., critical points of review and approval) and include a high-level timeline for the project. At a minimum, the project timeline should include the start and finish dates of each phase.

Major Milestones

Project Deliverables

Project Timeline

Out of Scope

Describe what is excluded from the scope of the project.

Assumptions

Describe any assumptions that this Project Charter is based on.

Acceptance Criteria

Answers the question, "How will we know if the project objectives have been met?"

Constraints

Describe any design constraints, external constraints or other dependencies. Note any pre-defined budget or schedule constraints.

Project Estimates

Provide the estimated costs in the table below. Following the Estimated Costs, describe the project benefits – tangible or intangible. The benefits stated should relate to the Acceptance Criteria defined earlier in this document. How costs are represented, vary from company to company and even from project to project but a good charter should have some measure of cost.

Estimated Costs

Project:	Year 1	Year 2	Year 3	Total
Expense - Labor				
Expense - Non-Labor				
Total Expense				
Capital - Labor				
Capital - Non-Labor				
Total Capital				
Grand Total				
On-going Operating Expense:				

Benefits

Direct Benefits

Indirect Benefits

Describe the project benefits – direct and indirect. An indirect benefit is a return that cannot be directly observed but is nonetheless realized — as opposed to direct benefits like reduced headcount or increased sales that are more easily quantified.

2.2   Status Report

Creating and circulating a Project Status document on a regular basis is a core project management activity; it allows you to keep staff on track, keep management informed of your progress and to seek guidance when needed.

The project status report is one of the most important documents created by the project management as it fulfills multiple critical purposes. The first purpose is to provide management and stakeholders with an accurate assessment of the progress of the project and to raise any items that may require their attention. The second purpose is to ensure that at least once during the reporting period, the project manager and team are forced to review due items and outstanding issues. As a likely result, the team is forced to focus on progress as well as resolving any outstanding issues.

The contents of the status report will vary depending on the size of the project, the organization culture, the diversity of stakeholders, the maturity of project management and many other factors, but ultimately it must communicate whether the project is on track and likely to finish within schedule.

The status report can include details on schedule, budget, staffing, deliverables, issues and risks.

Typically, project reports are distributed on a weekly basis to all stakeholders and can be distributed before or after a regularly scheduled status meeting.

 It is critical that the information in the project report is accurate and up to date. The report should only include information that is relevant to sponsors and stakeholders and should answer questions like: Are your expenses within budget? Is the project on schedule to meet predefined deadlines? Will any risks likely affect the project? Are any issues impacting the project?

The project manager should define the contents and format of the project report. A starting point is to look at what are the important things in the particular project at hand. These will vary from project to project but here are a few typical criteria for project reporting.

Schedule - How is the project progressing against the schedule.

Budget - How is the project progressing against the budget.

Risks - What risks have been identified and how are they being managed

Issues - What new issues have arisen and how are they being managed.

The status report should be an exception report and keep in mind that the point of the report is to get attention focused on the tasks that are not going according to plan. It should bring the spotlight onto the issues the project team wants the readers to focus on.

Status Report Sections

Schedule and Accomplishments
- List activities completed during the reporting period
- List activities that were not completed as planned
- List activities planned for next reporting period

Budget

In some organizations the budget should be reported in every status report; while, in others the budget should not be distributed to every person that receives the status report. You need to make a judgmental call and decide who should receive status reports on the budget and the frequency. It does not have to be part of the regular status report.

Issues and Risks

Issues must be assigned for resolution to a team member, or team members, and must have a deadline for resolution.

Once the project is underway and after the initial start up, issues and issue tracking can be the bulk of the status report.

Project Status

A good way to represent the overall status of a project is by using the standard colors red, yellow and green with the following meanings:

- Green (Controlled) – Project is within budget, scope and schedule. Green means that the project is mostly working as planned and that deadlines are going to be met.
- Yellow (Caution) – Project has deviated slightly from the plan.
- Red (Critical) – Project has fallen significantly behind schedule, it is forecasted to be significantly over budget, or has taken on tasks that are out of scope.

Format of Status Report

This is the template of a Status Report. The instructions are italicized text within gray shaded boxes and should not be included in the final document

Header Information

This is where you identify the entity doing the project, the entity that owns the project, the stakeholders, the project manager and the team members.

The header will also include the dates that the status report covers.

Status Report

Project Name:
Date:
Dates covered by Status report:

Overall Project Status

Project Status
This is where you detail the overall project status and where you use the colors described above.

Summary Status

Summary supporting the R/G/Y status.

Activities Performed on the Reporting Period

List the accomplishments for the last reporting period.

Issues and Risks

List the issues and risks that are open and assign an owner to each of them.

Activities for Next Reporting Period

List the activities to be performed during the next reporting period.

Calendar items

List next meetings, deadlines, travel of team members, etc.

Budget

The report can include a section that shows the total budget and the budget used to date. A very valuable indicator to include in these cases is the "burn rate". The burn rate is the cost per week of the project and of course, this is meaningful if the status also includes the estimate of weeks left.

Progress Chart

This is a bar chart that shows the progress of major activities and the milestones.

The dark areas reflect progress and white areas represent planned work. This bar chart reflects that the design task was completed as scheduled, but that the development task is behind. On the other hand testing is ahead of schedule.

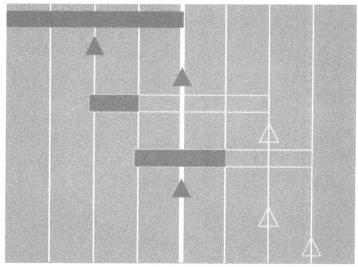

Activity/Milestone	Progress report as of the end of Week 4
2.0. Design	
2.2.4. Initial design completed	
2.3.6. Final design completed	
3.0. Develop	
3.4. Development completed	
4.0. Test	
4.1.4. Test plan completed	
4.3.6. Initial resting completed	
4.6.7. Final resting completed	

Week 1   Week 2   Week 3   Week 4   Week 5   Week 6   Week 7   Week 8

Earned Value

The report could include an earned value analysis, but each project should determine if the effort to do the analysis is justified for the project.

PMP Certification and PMP Test Exercises

Project Management Institute®

3.1　The PMP Certification Test

The Project Management Institute (PMI) certifies project management professionals and confers the PMP® designation. The PMP® designation offers individuals worldwide recognition as professionals who have received project management certification, supported by core knowledge and understanding of critical project management concepts, principles and techniques. This is an impressive certification and one that every fulltime project manager should aim to achieve.

PMI has found that project managers that have the PMP certification earn up to 10% more than non PMPs in the United States.

The PMP certification requires heavy learning and actual real life practice with the examination covering real life project management and project management theory as defined by PMI. This book concentrates on the most practical aspects of project management. It covers what a project manager absolutely must know in order to do his or her job, but it does not cover all the topics that are included in the PMP examinations, as that is not its focus. Nevertheless, we have included topics dealing with the project management certification test. Those topics are: how to calculate a critical path in a network diagram and exercises on calculating earned value.

While this book describes every process defined by the PMI, it does not include the detailed information about inputs, outputs or documents related to each process. The PMP certification test asks about those processes in detail and you will have to read the "Guide to Project Management Body of Knowledge" in order to be able to answer those questions.

Advise for taking the PMP examination:
- The PMP certification has been changing over the years, so make sure you understand the current requirements and consult reference material relevant to the current test.
- The PMP certification is not simple a test of project management theory (like the one you acquire with this book) but also real life situational cases. It is important to know how to apply the theory to real life.
- Many questions are about selecting the best course of action even when two or more of the suggested answers are valid but they are not the best.
- About 10% of the questions are mathematical or procedural. They will involve questions about finding the critical path and float using the forward and backward passes, and earned valued formulas. (This book does cover these areas.)
- Many questions are based on the PMI - defined project management knowledge areas and processes.
- You must read the current Project Management Body of Knowledge (PMBOK) from PMI. Keep in mind that that book is not intended to actually teach Project Management Body of Knowledge, but describe it. You will need additional material, like this book, to understand in full the knowledge required to pass the certification.

3.2   Critical Path Calculation

Most project managers use some kind of software to help identify the critical path and the duration of the project, but for the PMP certification test, you will need to know about network diagrams and know how to calculate the critical path of a project.

The process to identify the critical path involves doing a "forward pass" and a "backward pass" in a network diagram, and while all this is done for you by project manager systems you do need to understand how it works.

Network Diagram

The first step in calculating the critical path is listing all the activities that must be performed to complete the project. The next step is to identify the sequence, or order in which those tasks should be executed. As the following diagram shows, this can be done with a network diagram which shows the tasks and their relationships of precedence.

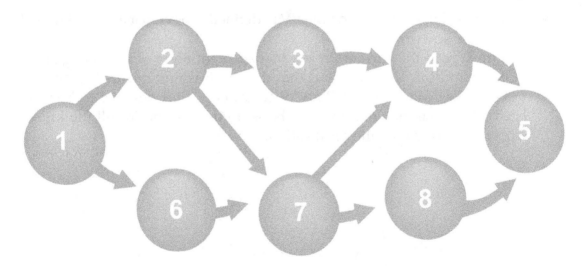

In this network diagram, the tasks are represented by the circles and the sequence or order of execution is represented by the arrows. For example, an arrow going from task 1 to task 2 means that the task 1 must be completed before task 2 can be completed.

To calculate floats and critical paths you need to start by creating a precedence diagram. A precedence diagram is nothing else than an activity on node diagram and is built following the next steps:

1. Draw a start node and draw the activities that do not have predecessors.
2. Draw arrows and activities to following tasks.
3. Continue until all activities and predecessor relationships have been represented.
4. Draw end node and predecessors to it.

In a network diagram each activate duration is represented with a box like this one:

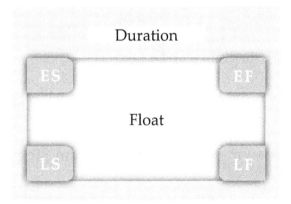

Where:

Duration	=	Estimated duration
ES	=	Early Start
LS	=	Late Start
LF	=	Late Finish
EF	=	Early Finish
Float	=	Calculated float

- Early Start - Is the earliest date that the task can start.
- Late Start - Is how late it can start without delaying the project.
- Early Finish - Is the earliest date that the task can be completed.
- Late Finish - Is how late the task can finish without delaying the project.

Usually, all those dates are set as a number, instead of an actual date, where 1 is the "date" when the first task (or tasks) will start.

Example of Drawing of a Network Diagram

This example shows how to build a network diagram with activity on node. This same example will be used later for the calculation of float and critical path.

Task	Duration	Predecessor(s)
A	3	None
B	5	None
C	4	A
D	2	C
E	4	B

Example 1.	Task List

Drawing of network diagram step by step

1. Draw a start node and the activities that do not have predecessors.

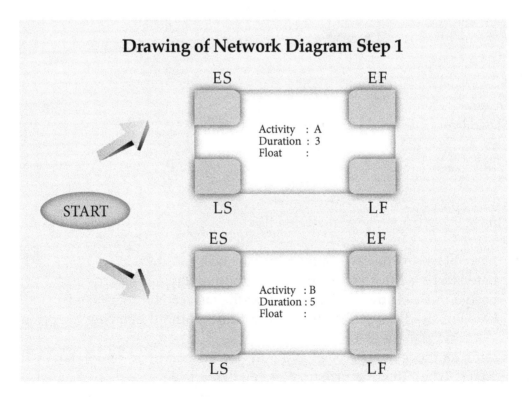

2. Draw arrows and activities to following tasks.

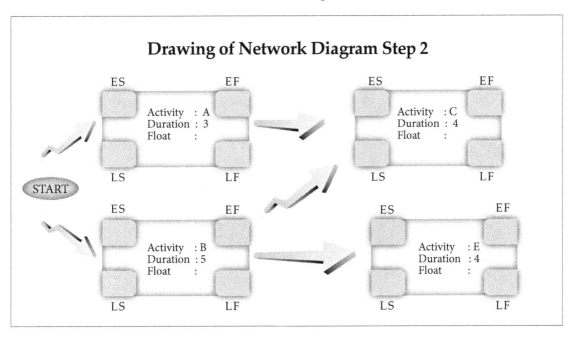

Drawing of Network Diagram Step 2

3. Continue until all activities and predecessor relationships have been represented.

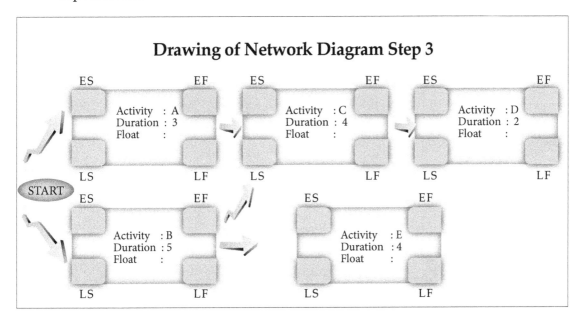

Drawing of Network Diagram Step 3

4. Draw end node and predecessors to it.

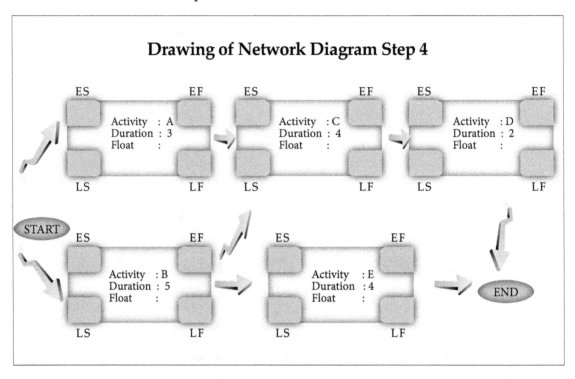

Drawing of Network Diagram Step 4

Calculation of critical path

The first part of the calculation is the "forward pass" and it starts by putting a "0" (zero) on the Early Start box of the activities that don't have predecessors. This "0" represents the first day in the project calendar (the durations for this example are given in days).

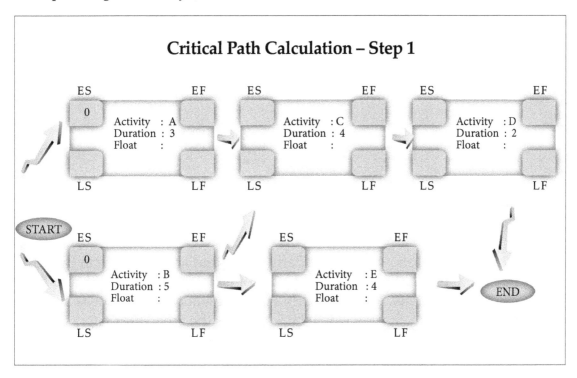

Then, add the duration of the activities to the Early Start and put the result in the Early Finish box.

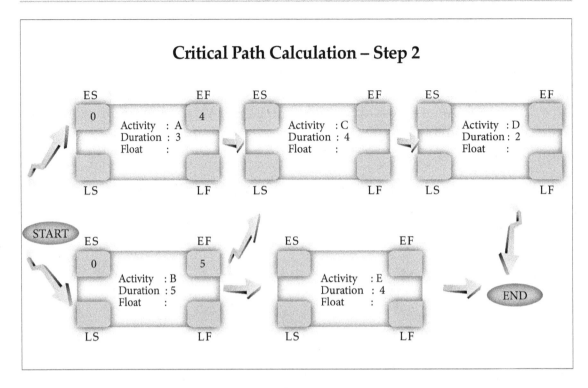

Critical Path Calculation – Step 2

The activities that follow will have Early Starts that are greater than the Early Finish dates of the activities that precede it. For activity C, activities A and B are predecessors and activity A has an Early Finish of 4 and activity B has an Early Finish of 5, so the Early Start of C will be 5 which is the greater (or later) than the Early Finish of the predecessors. Activity E has only one predecessor and its Early Start will just be the Early Finish of its unique predecessor B.

Critical Path Calculation – Step 3

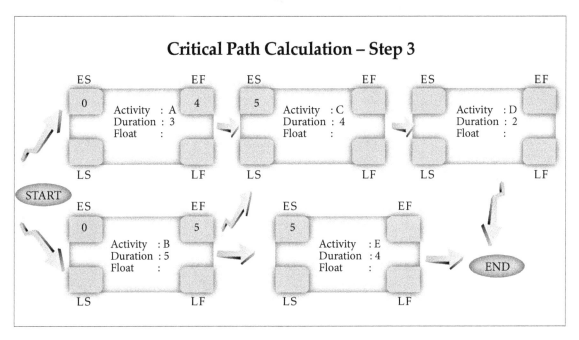

Critical Path Calculation – Step 4

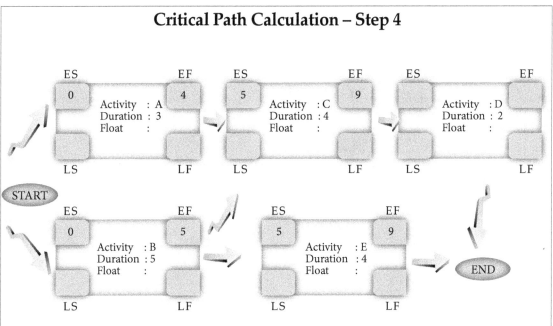

Add the duration of the activities to find out the Early Finish dates.

Continue until all activities have been analyzed.

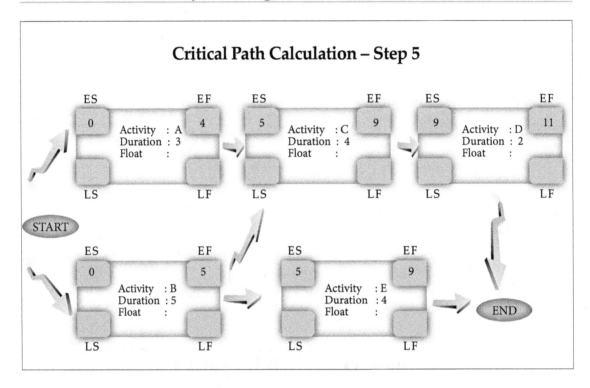

Now you can find out the Early Finish of the project by looking at the greatest of the Early Finish dates for all activities that are predecessors of the end node. In this case, the greatest of 9 and 11 is 11 and, therefore, the Early Finish date for the project is 11.

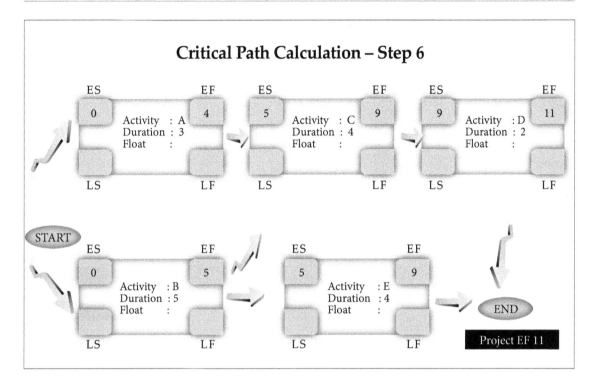

Critical Path Calculation – Step 6

Now that we know how soon the project can end, we will look for the critical path doing a backward pass.

The first step of the backward pass is to enter the Late Finish dates of the activities that precede the End Node.

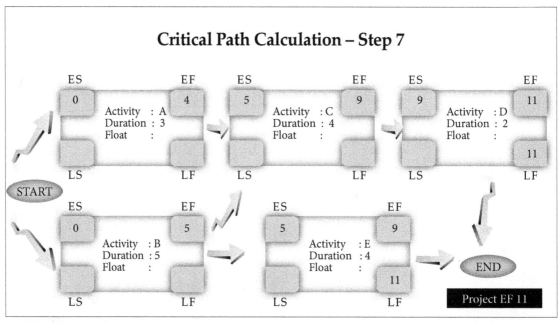

We then subtract the Duration form the Late Finish to find out the Late Start dates.

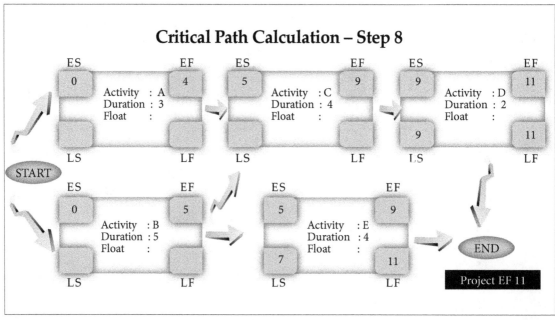

We then proceed backwards to find all the Late Start and Late Finish dates. For tasks that have a single successor (like tasks C, D and E) their Late Finish dates are just the Late Start date of the successor as shown for tasks C, D and E.

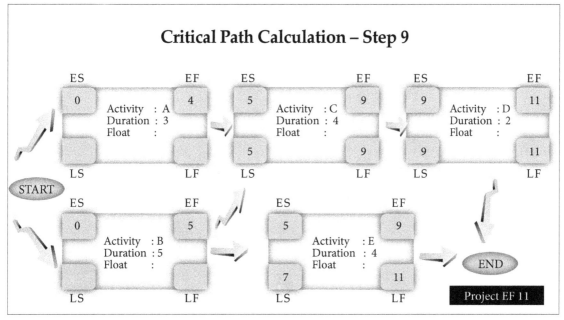

For tasks that have more than one successor (like task B), the Late Finish date, will be the lesser of the Late Starts of their Successors. In this case, the smallest of 5 and 7, which is 5.

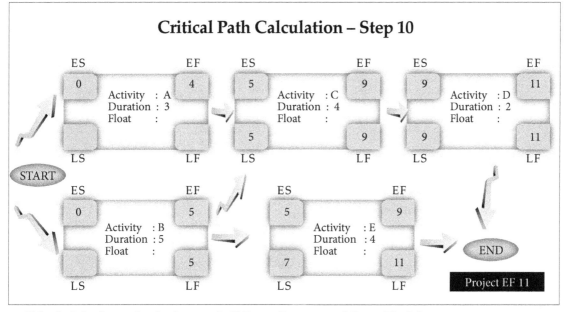

We finish the calculations of all Late Starts and Late Finishes.

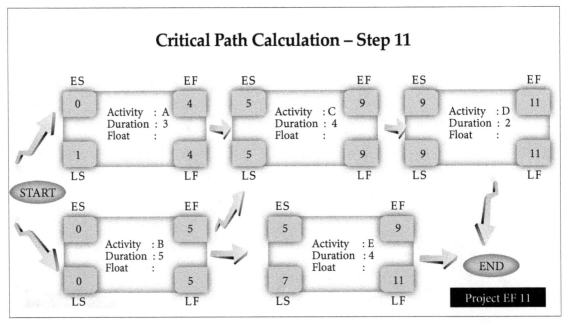

Now we can calculate the Float by using either of two formulas for Float: LF - EF or LS – ES.

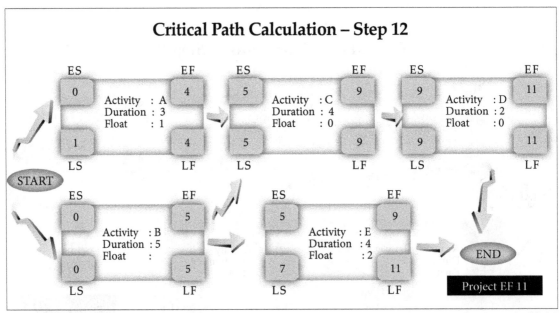

The activities with Float equal to zero are in the critical path marked with bold lines below:

Critical Path Calculation – Step 13

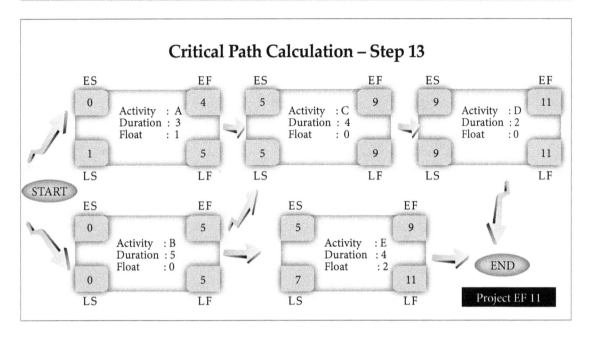

Another example:

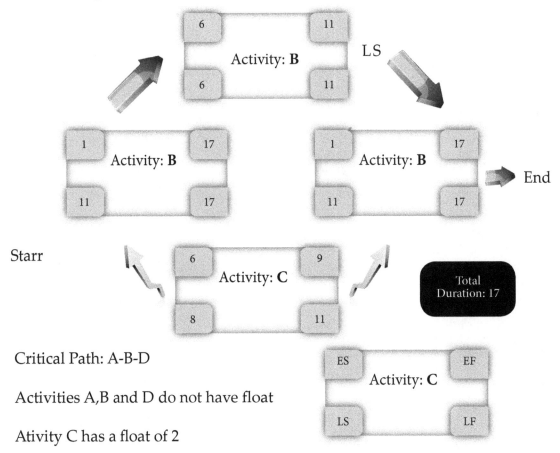

Starr

Critical Path: A-B-D

Activities A,B and D do not have float

Ativity C has a float of 2

The list of tasks and floats looks like this:

Task	Duration	Predecessor(s)	Float	Critical Path
A	3	None	1	No
B	5	None	0	Yes
C	4	A,C	0	Yes
D	2	C	0	Yes
E	4	B	2	No
Example 1.		Float and Critical path activities		

3.3 Earned Value Examples

Tasks

Task	Duration	Predecessor(s)	Value
A	3	None	$ 100.00
B	5	None	$ 300.00
C	4	A,C	$ 200.00
D	2	C	$ 100.00
E	4	B	$ 200.00

Example 1.	Float and Critical path activities

Task	Day 1	Day 2	Day 3	Day 4	Day 5	Day 6	Actual Cost	Status at end of day 6
A							$100	Completed
B							$400	Completed
C							$100	20%
D								
E							$800	10%

Shaded lines are planned work.

Review the following values and try replicating the values of the last four columns by using the following Earned Value formulas:

Plan Value - Plan value is the dollar amount of progress that should be completed by a given day.

In the example, we are figuring out values at the end of day 6. The planned values are calculated under the assumption that the derived working task value is lineal with the actual days worked on the task. . So for instance, task C has a duration of 4 days, and at the end of day 6, there has been a full day of work on the task, so the planned value should be 1/4th of the total value. The planned value, then, for task C is 1/4th of $, and that is equal to $50.

Earned Value - Earned value is calculated multiplying the % complete against the budget for the task.

Schedule Variance -Schedule Variance is equal to Earned Value - Planned Value.

Cost Variance - Cost Variance is equal to Earned Value - Actual Cost.

Schedule Performance Indicator - Earned Value / Planned Value

Cost Performance Indicator - Earned Value / Actual Cost

Task	Total Duration	Budget	% Comp.	Planned Work Days to day 6	% Plan Value	Plan Value	Earned value	Actual Cost	Schedule Variance	Cost Variance	Schedule Performance Indicator	Cost Performance Indicator
A	3	$100	100%	3	100%	$100	$100	$100	$0	$0	1,00	1,00
B	5	$300	100%	5	100%	$300	$300	$350	$0	-$50	1,00	0,86
C	4	$200	20%	1	25%	$50	$40	$100	-$10	-$60	0,80	0,40
D	2	$100	0%	0	0%	$0	$0	$0	$0	$0		
E	4	$200	10%	1	25%	$50	$20	$80	-$30	-$60	0,40	0,25
Total		$900				$500	$460	$630	-$40	-$170	0,92	0,73

Overall Project Values

Earned Value Indicators	Value
Plan Value	500
Earned value	460
Actual Cost	630
Schedule Variance	-40
Cost Variance	-170
Schedule Performance Indicator (SPI)	0.92
Cost Performance Indicator (CPI)	0.73

These are some conclusions that can be extracted:
- The project is behind schedule. The SPI says that for each dollar planned the project only accomplished 92 cents.
- The project is above budget. The CPI says that for each dollar expend, the project only obtained a value of 73 cents.

Remember that when analyzing the Schedule and Cost Variances, negative numbers are bad, positive numbers are good. Similarly, SPI and CPI values below one are bad, and values above one are good.

In this case, the Schedule Performance indicator is .92 meaning the project is below schedule but a .92 might not be too bad of a performance. The Cost Performance Indicator, in the other hand, is just .73 meaning that the cost of the project is higher than expected.

Introduction to Microsoft Project

Microsoft Project

Microsoft Project (MSP) has many options and features, and for a person who has never seen it, it might be intimidating. This tutorial will show you enough to create a simple project and track it. Once you complete the tutorial you will be able to manage projects in MSP and be ready to start exploring all its features on your own.

Examples are made with MSP 2010 Professional. If you have another version, your screen and options may look different

Microsoft Project Software

The graphics included in this section are simple screen shots that do not work well in print. They are just a guide to what you can actually see in MS project. In order to learn the basics of MicroSoft project with this book and to follow the examples, you will need to have access to Microsoft Project software. If you don't have it installed, you can download, for free, a trial version from Microsoft. This following link will take you the MicroSoft website from where you can download the software for free: http://office.microsoft.com/

4.1 Starting a New Project

Open MSP and you will see a screen like this:

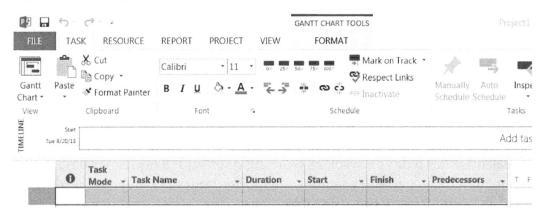

Figure 27 - MS Details of Upper Right Corner MSP First Screen

When you have a new project, you may be tempted to start entering your tasks, but the very first thing you need to do is set up the "project information" for your project. To do so go to 'File', select 'New' and then select 'Blank Project'. Go to the 'Project' tab and select "Project Information" This will open the project information window.

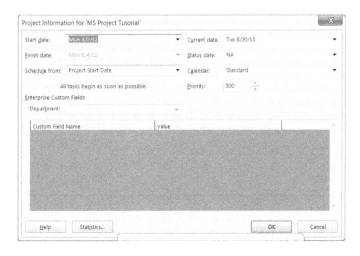

Figure 28 - Project Information screen

4.2   MSP Project Information

Enter the project start date and set the "Schedule from" field to "Project Start Date". You could also enter the finish date, but that is used when you must finish by a certain date and you do the task scheduling backwards. For a first time project, let us stick to   scheduling forwards. Since it takes a few days to compile and begin executing a plan, it is better to enter a starting date ahead of the current date. Otherwise, the tasks you enter will appear late even if the project has not been yet started. Alternatively, make your first task a planning task.

The 'Project Information' window allows you to set up a calendar. The standard calendar has 5 days a week as working days with 8 hours per day. You can make many changes to the calendar, including hours per day and adding holidays. Click 'OK' to save the selections.

It is convenient to save your work from time to time so let's start by saving the project with the project information. Go to 'File' tab and then select 'Save As'.

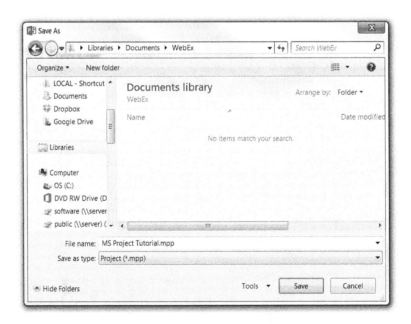

Figure 29 - Save Dialog screen

Enter a file name and click save.

4.3 Entering Tasks

Go to the 'Task' tab and select 'Gantt Chart'. This is the screen where you will enter the project activities and data like task name, duration, start date predecessors and other info. You can enter the required information in two ways: in a spreadsheet format or in a pop up window that you get when you double click on a cell. For a predecessor activity, enter the activity number.

In the first line of the spreadsheet and in the column "Task Name" enter the name of the project. Then enter the activities that will be executed in order to complete the project.

Select all activities below the name of the project and make sure they are indented to the right. To "indent" an activity select it, do a right click and select the right arrow as shown below:

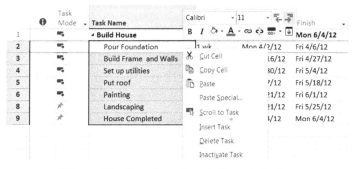

Figure 30 - *How to make a task a sub-task*

Enter the duration of the activities:

		Task Mode	Task Name	Duration
1			⊿ **Build House**	**45 days**
2			Pour Foundation	1 wk
3			Build Frame and Walls	2 wks
4			Set up utilities	1 wk
5			Put roof	2 wks
6			Painting	2 wks
8			Landscaping	1 wk
9			House Completed	0 days

Figure 31 - *Activity duration data entry*

4.4 Order of Execution

Now we are going to enter the predecessors. Move the sliding window to see the predecessors' column and enter the predecessor of each activity. The predecessor is entered by typing the activity number of the predecessor activity in the predecessor column:

		Task Mode	Task Name	Duration
1			◢ **Build House**	**45 days**
2			Pour Foundation	1 wk
3			Build Frame and Walls	2 wks
4			Set up utilities	1 wk
5			Put roof	2 wks
6			Painting	2 wks
8			Landscaping	1 wk
9			House Completed	0 days

Figure 32 - *Tasks and their predecessors*

Note that the Task Mode column has pins on it and that means that the task is in manual schedule. We are going to change them to auto-schedule. The auto-schedule option will make the start dates change as the end dates of the predecessors' change.

Select all the tasks, do a right click and select auto schedule:

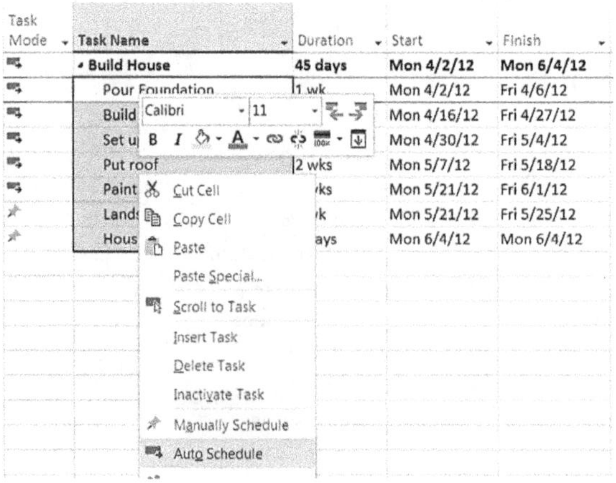

Task Mode	Task Name	Duration	Start	Finish
	◢ Build House	45 days	Mon 4/2/12	Mon 6/4/12
	Pour Foundation	1 wk	Mon 4/2/12	Fri 4/6/12
	Build		Mon 4/16/12	Fri 4/27/12
	Set u		Mon 4/30/12	Fri 5/4/12
	Put roof	2 wks	Mon 5/7/12	Fri 5/18/12
	Paint	rks	Mon 5/21/12	Fri 6/1/12
	Lands	rk	Mon 5/21/12	Fri 5/25/12
	Hous	ays	Mon 6/4/12	Mon 6/4/12

Cut Cell
Copy Cell
Paste
Paste Special...
Scroll to Task
Insert Task
Delete Task
Inactivate Task
Manually Schedule
Auto Schedule

Figure 33 - *Auto Schedule is the preferred option for scheduling tasks*

You should see that all tasks have changed to auto schedule:

	ⓘ	Task Mode ▾	Task Name ▾	Duration ▾	Start ▾	Finish ▾	Predecessors ▾	Free Slack ▾
1		🖥	⊿ Build House	45 days	Mon 4/2/12	Mon 6/4/12		0 days
2		🖥	Pour Foundation	1 wk	Mon 4/2/12	Fri 4/6/12		0 wks
3		🖥	Build Frame and Walls	2 wks	Mon 4/16/12	Fri 4/27/12	2FS+1 wk	0 wks
4		🖥	Set up utilities	1 wk	Mon 4/30/12	Fri 5/4/12	3	0 wks
5		🖥	Put roof	2 wks	Mon 5/7/12	Fri 5/18/12	4	0 wks
6		🖥	Painting	2 wks	Mon 5/21/12	Fri 6/1/12	5	0 wks
8		✯	Landscaping	1 wk	Mon 5/21/12	Fri 5/25/12	5	1 wk
9		✯	House Completed	0 days	Mon 6/4/12	Mon 6/4/12	7,6	0 days

Figure 34- Tasks change their starting dates based on the ending dates of predecessors

Normally, after the foundation is poured, the cement needs some time to cure. This delay is represented as follows:

Move your cursor to the "Build Frame" task and click twice with the left button on your mouse. The 'Task Information' window will appear. Click on the 'Predecessors' tab, go to the column 'Lag' and enter "1 w "(one week).

Figure 35 - Entering lag or wait time for a task

This is telling MSP that the task can start 1 week after the predecessor task has finished.

To see the tasks in a graphic manner go to the chart section of the screen on the right and right click on the calendar section. You should see a menu option with a time scale:

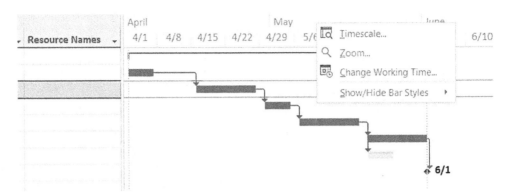

Figure 36 - MS Project change of time scale

Select the 'Timescale' option and the 'Timescale window' will appear. Click on the 'Middle Tier' tab and make sure that the options are set up as follows:

Units: Months,

Timescale options - Show: Two Tiers (Middle, Bottom)

Figure 37 - Time Scale options

Right click on the "Bottom Tier" tab and make sure the units are in weeks. Click 'OK'.

You should be able to see the Gantt chart for the project:

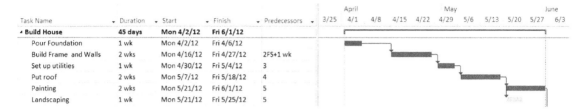

Task Name	Duration	Start	Finish	Predecessors	April			May			June
⁀ Build House	45 days	Mon 4/2/12	Fri 6/1/12								
Pour Foundation	1 wk	Mon 4/2/12	Fri 4/6/12								
Build Frame and Walls	2 wks	Mon 4/16/12	Fri 4/27/12	2FS+1 wk							
Set up utilities	1 wk	Mon 4/30/12	Fri 5/4/12	3							
Put roof	2 wks	Mon 5/7/12	Fri 5/18/12	4							
Painting	2 wks	Mon 5/21/12	Fri 6/1/12	5							
Landscaping	1 wk	Mon 5/21/12	Fri 5/25/12	5							

Figure 38 - Typical Gantt chart

Note how the tasks of painting and landscaping can be done at the same time.

You can add tasks at any time. For example, we will add the task that represents the culmination of the project. This task is going to be a milestone.

Go to the 'View' tab and select 'Gantt Chart'.

Put your cursor on the first empty row and right click twice. The task information screen will appear.

Enter the name of the milestone, "House Completed". Duration of zero is used to represent a milestone so enter a zero in 'duration':

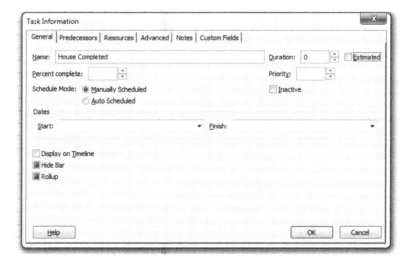

Figure 39 - Task Information

Then, go to the 'Predecessors' tab and type a 7 in the ID column, hit tab and then type a 6 in the next row:

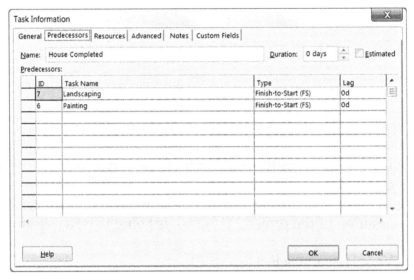

Figure 40 - Task Information - Predecessors

Click 'OK'.
You should see a diamond at the end of the Gantt chart:

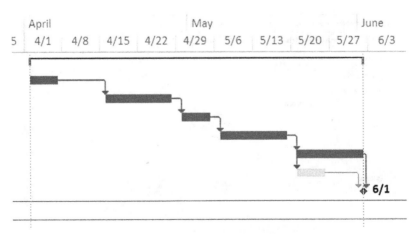

Figure 41 - A diamond represents a milestone

The diamond is the representation of a milestone.

4.5 Critical Path

CPA helps you identify the minimum length of time needed to complete a project, and you can use MSP to do that for you. You can see the critical path in multiple ways: in the Gantt chart, in a network diagram and filtering on the critical tasks.

In the Gantt chart view, click on Gantt Chart Tools tab and then click on the Critical Tasks box. That will highlight the tasks on the Critical Path:

Figure 42 - Critical Path

Another way to see the critical path tasks are by filtering the critical tasks. Go to the View tab and then select Filter on Critical Tasks:

Figure 43 - Tasks can be filtered

This will show only the critical tasks:

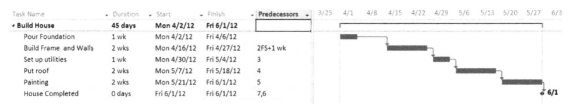

Figure 44 - Gantt chart showing only critical tasks

4.6   Slack Time

To view the schedule showing the slack go to 'Views', 'Other views', 'More Views', select 'Detail Gantt' view and click on 'Apply'.

In this view, the slack appears as thin bars to the right of a task, with slack values adjoining the regular Gantt bars. (In the example, you may have to move the cursor to the right to find the one activity that has slack.)

You can also view the free slack and total slack of a task in the sheet. Go to 'sheet', right click on the 'Predecessors' column and select "Insert Column". Finally, insert the column "Free Slack".

		Task Mode	Task Name	Duration	Start	Finish	Predecessors	Free Slack
1			Build House	45 days	Mon 4/2/12	Mon 6/4/12		0 days
2			Pour Foundation	1 wk	Mon 4/2/12	Fri 4/6/12		0 wks
3			Build Frame and Walls	2 wks	Mon 4/16/12	Fri 4/27/12	2FS+1 wk	0 wks
4			Set up utilities	1 wk	Mon 4/30/12	Fri 5/4/12	3	0 wks
5			Put roof	2 wks	Mon 5/7/12	Fri 5/18/12	4	0 wks
6			Painting	2 wks	Mon 5/21/12	Fri 6/1/12	5	0 wks
8			Landscaping	1 wk	Mon 5/21/12	Fri 5/25/12	5	1 wk
9			House Completed	0 days	Mon 6/4/12	Mon 6/4/12	7,6	0 days

Figure 45 - *Slack time representation*

You can move the activity within the available slack time.

4.7   Resources

You can use the 'Resource Sheet' in MSP to create a list of the people, equipment and material resources that make up your team and are needed to carry out the project tasks. Your resource list will consist of work resources or material resources. Work resources are people or equipment; material resources are supplies, such as concrete and wood.

Go to the view tab and then click on 'Resource Sheet'. You should see a screen like this:

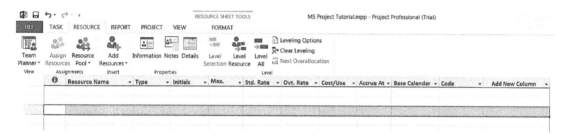

Figure 46 - Resource entry screen

Enter the resources names as follows:

	ⓘ	Resource Name	Type	Initials	Max.	Std. Rate	Ovt. Rate	Cost/Use	Accrue At	Base Calendar
1		Super Intendent	Work	PM	100%	$100.00/hr	$0.00/hr	$0.00	Prorated	Standard
2		Labor	Work	L	500%	$50.00/hr	$0.00/hr	$0.00	Prorated	Standard
3		Concrete Machine	Material	C		$0.00		$500.00	End	

Figure 47 - Example of resource entry

Note that there are not only personnel listed on the resource sheet but also a machine that will be rented for a particular job.

To assign resources to the tasks go back to the 'View' tab and 'Gantt Chart View'. Depending where you are, you may have to go to 'View', 'Other Views', 'More Views' and select 'Gantt Chart'.

In the 'Gantt Chart' view, make sure you can see the 'Resource names' column; for that, you may have to slide the dividing line between the sheet and the graphical portion of the Gantt chart.

You can enter resources directly in the 'Resources' column but, first, let's do it in the Task Information window. Select the Task ID 2 and right click twice on it to open the 'Task Information Window'.

In this window, select the resources tab and in the "Resource Name" column select "Super Intendent". Leave the units assigned at 100%. Enter a second Resource and this time select 'Labor' but this time enter 400% units. What this means is that four people will work on the task. Finally, enter the last resource, which is the Concrete Mixer with one unit.

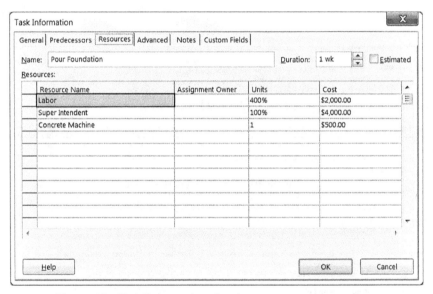

Figure 48 - Task Information showing resource usage and cost

4.8 The MSP Scheduling Formula

MSP is very flexible and allows for diverse levels of planning. While many people only plan tasks and duration as that is enough for some projects, other projects require planning for the resources (people) available. To handle this, MSP uses the project management concepts of resources, duration and work.

Resources are the people, equipment and supplies used to complete tasks in a project. Resource units indicate how much of a resource's available time, according to the resource calendar, is being used to work on a particular task. In a project schedule, resource effort is expressed as assignment units or sometimes just as units.

Duration is the span of elapsed time required to complete a task. (This is measured in hours, days or weeks.)

Work is the total effort required to complete a task and is defined as:

Work = Resources * Duration

MSP allows for the user the make any of these elements fixed. By fixing one of the elements, when one of the other elements is changed by the user for a given task, the system recalculates the non-fixed element to preserve the formula as detailed below.

Fixed Units
Fixed units are the default task type for new tasks. As the task's assigned units remain constant, an increase/decrease in assigned work will result in a corresponding increase/decrease in task duration.

Fixed Work
If you set the task to have a fixed work value, once you enter the work value of the task it remains fixed. If more units are added, the duration will be reduced. On the other hand, if the duration is increased, the units will be reduced.

If the calculated or assigned units are greater than the maximum units available, a resource over allocation will be detected by MSP.

Fixed Duration

For a constant task duration, as assigned work is increased so will assigned units. Conversely, decrease the units and the work will correspondingly reduce.

If assigned units are greater than max units, a resource over allocation will arise.

Effort-Driven Tasks

Effort driven is the default option for new tasks. When a task is marked as 'Effort-Driven', MSP keeps the work constant, and when resources are assigned or removed from a task, MSP will extend or shorten the duration of the task to accommodate change in resources. "Effort driven" tasks are those whose duration truly reduces if more resources are applied to it. For instance, the time it takes to paint a house by one painter could be reduced to half of it, if we add another painter to the job.

Beginner users of MSP should start with fixed duration tasks and move to effort driven tasks as they gain experience with the system.

Appendix

***Figure 49** - The International Space Station - One of the most expensive projects in history*

5.1 Project Management Historical Milestones

1910's - Henry Gantt invents the Gantt Chart.

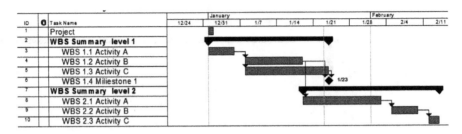

1950's - Du Pont Corporation and Remington Rand Develop the Critical Path method.

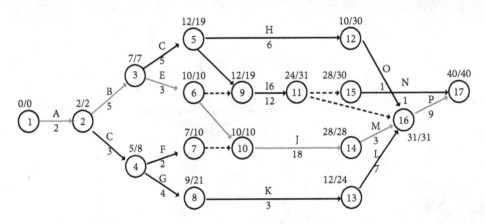

1950's - Booz-Allen & Hamilton and The United States Navy develop PERT.

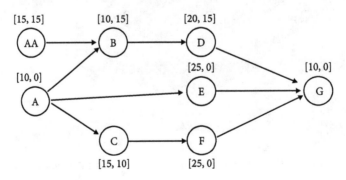

1956 - The American Association of Cost Engineers (now AACE International) is founded.

1969 - The Project Management Institute is founded.

1987 - A Guide to the Project Management Body of Knowledge (PMBOK) is first published.

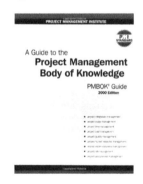

2000, 2006, 2008 - Second, third, and fourth editions of the PMBOK are published.

- 2012 - Close to half a million people have received the PMP certification.

5.2   Project Management Formulas

Float and Critical Path
There are two formulas for float:
Float = LS - ES
Float = LF - EF
Where LS = Late Start; ES = Early Start; LF = Late Finish; EF = Early Finish

PERT
Activity Duration = (O + 4 * M + P) / 6
Where O = Optimistic Estimate; M = Most Likely Estimate; P = Pessimistic Estimate;
Activity Standard deviation = pessimistic estimate –optimistic estimate/6
One standard deviation = 68 +/- % on a bell curve
Two standard deviations = 95 +/- % on a bell curve
Three standard deviations = 99 +/- % on a bell curve

Risk
Expected Monetary Value (EVM) = Risk event probability * event value
Expected Cost = Sum of Expected Monetary Value
Communications
The number of communications channels = n (n-1)/2 where n is the number of people.

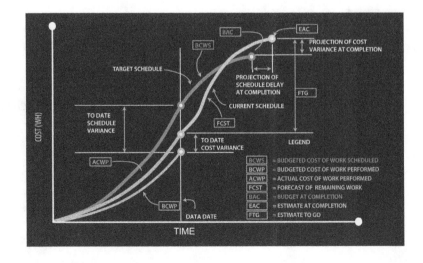

Earned value acronyms: Earned value = EV, Planned Value = PV, Actual Cost = AC.

Remember that in Earned Value formulas the value of Earned Value always comes first and that you subtract for variance and divide for performance.

Schedule variance: $SV = EV - PV$ (positive means ahead of schedule, negative means behind schedule)

Cost Variance: $CV = EV - AC$ (positive means under budget, negative means over budget)

Cost Performance Indicator: $CPI = EV/AC$ (Over 1 is good)

Cumulative CPI = sum of EV /sum of AC

Schedule Performance Indicator: EV/PV (Over 1 is good)

Variance at completion $VAC = BAC - EAC$

Where BAC = Budget at Completion and EAC = Estimate at Completion.

There are four ways to calculate EAC:

- $EAC = BAC/CPI$ (assumes that whatever affected past results will also affect future results)
- $EAC = AC$ + remaining BAC/CPI (alternate way to get same results)
- $EAC = AC$ + remaining BAC (assumes that whatever affected past results will NOT affect future results)
- $EAC = AC + ETC$ (assumes that enough has changed to warrant a new estimate)

5.3 Internet Resources on Project Management

American Academy of Project Management

http://www.projectmanagementcertification.org/

American Society for the Advancement of Project Management

http://www.asapm.org/

International Project Management Association

http://www.ipma.ch/

PM Forum

http://www.pmforum.org/

Project Management Institute

http://www.pmi.org/

About the Author

George T. Edwards obtained the Project Management Professional certification in 2003 and has over 20 years of project management leadership. Mr. Edwards has managed projects in the Americas and Europe for top global companies for Fortune 1000 US firms

Mr. Edwards has worked for a Big 5 Management consulting firm, for major software companies and as an independent project management consultant. While most of his experience as a project manager has been in the information technology area, he also has experience with civil engineering projects.

Mr. Edwards holds degrees in Industrial Engineering, Information Systems and Computer Science, and he holds information technology related certifications.

www.ingramcontent.com/pod-product-compliance
Lightning Source LLC
Chambersburg PA
CBHW080421060326
40689CB00019B/4327